DATE DUE

SCIENTISTS AT WORK

Scientists at Work

THE CREATIVE PROCESS OF SCIENTIFIC RESEARCH

Edited by
JOHN NOBLE WILFORD

Illustrated with photographs

DODD, MEAD & COMPANY · NEW YORK

Printed in the United States of America

1 2 3 4 5 6 7 8 9 10

Library of Congress Cataloging in Publication Data

Main entry under title:

Scientists at work.

Originally published as articles in the New York Times.
Includes index.
1. Research—History—Collected works. 2. Science—History—Collected works. I. Wilford, John Noble.
Q180.A3S37 508'.1 78-18256
ISBN 0-396-07603-3

CONTENTS

Introduction by John Noble Wilford xi

I. In Search of the Past
ARCHAEOLOGY, PALEONTOLOGY, DENDROCHRONOLOGY

1. *Autopsy on a Mummy* by Boyce Rensberger. August 28, 1975* 7
2. *Time and the Bristlecone Pine* by John Noble Wilford. December 27, 1975 13
3. *A New Look at Old Skulls* by Boyce Rensberger. April 21, 1976 20
4. *A Revised Look at Man's Entry into the New World* by Boyce Rensberger. August 16, 1976 27
5. *Clues to the Life of Stone-Age Man* by Boyce Rensberger. July 28, 1977 36

II. *Earth, Sky, and the Subatomic World*
ASTROPHYSICS, GEOLOGY, PHYSICS

6. *Explorers in a Subatomic World* by Walter Sullivan. July 12, 1975 49
7. *Predicting Earth's Most Violent Tendencies* by John Noble Wilford. May 18, 1977 61
8. *An Astronomer Looks Back to See Ahead* by Walter Sullivan. November 28, 1977 68

III. *Invention*
ELECTRONICS

9. *The Ferreed Crosspoint: A Revolution in Telephones* by Victor K. McElheny. October 25, 1975 79

*Date of publication in *The New York Times*

IV. *Exploration of Life—I*
BIOLOGY, CHEMISTRY, GENECTICS, PHYSIOLOGY

10. *The Chemistry of Sex and Pest Control* by Jane E. Brody. May 25, 1975 95
11. *To Breed a Perfect Elm* by Bayard Webster. November 29, 1976 103
12. *A Tomato That Can Stand the Cold* by Jane E. Brody. May 14, 1977 110
13. *Genetic Decoders* by Walter Sullivan. June 20, 1977 117
14. *A Strange Bacterium's Purple Pigment* by Jane E. Brody. July 5, 1977 129
15. *Uncovering Secrets of Insect Life* by Boyce Rensberger. October 18, 1977 137
16. *A Jungle Field Station* by Walter Sullivan. November 3, 1977 144
17. *Bears: A Search for the Sleep Hormone* by Lawrence K. Altman. November 21, 1977 149
18. *Unraveling the Shape of an Enzyme* by Malcom W. Browne. December 7, 1977 155
19. *A Search for a New Male Contraceptive* by Jane E. Brody. February 21, 1978 161

V. *Exploration of Life—II*
ANIMAL BEHAVIOR, ECOLOGY, LIMNOLOGY

20. *A Forest as a Living Lab* by Bayard Webster. June 5, 1975 173
21. *Wading in the Name of Science* by Bayard Webster. June 4, 1976 180
22. *A Better Image for the Wolf* by Boyce Rensberger. February 16, 1977 186
23. *Rats: Living Monitors of Radiation* by John Noble Wilford. April 18, 1977 193
24. *The Countless Mysteries of Peatland* by Jane E. Brody. October 6, 1977 201

VI. *Science and Health*

BACTERIOLOGY, NUTRITION, VIROLOGY

25. *Fat and Life-Saving Clues* by Harold M. Schmeck, Jr. September 2, 1975 213
26. *The Race for a Swine Flu Vaccine* by Harold M. Schmeck, Jr. May 21, 1976 221
27. *The Key to Legionnaires' Disease* by Harold M. Schmeck, Jr. January 24, 1977 229
28. *Penicillin-Resistant Gonorrhea* by Harold M. Schmeck, Jr. June 29, 1977 237
29. *A New Method of Intravenous Feeding Saves Lives* by Jane E. Brody. November 25, 1977 244

Biographies of Authors 255
Index 261

ILLUSTRATIONS

Dr. Theodore Reyman and Dr. Michael Zimmerman perform an autopsy on an Egyptian mummy 3
Dr. Henry N. Michael examining the base of a bristlecone pine 4
Ralph L. Holloway, Jr., measuring the skull of an extinct ancestor of modern man 4
Dr. Richard S. MacNeish and students examining material taken from a dig 5
Archeological students on a dig in western Pennsylvania 6
Apparatus at Stanford in which positron-electron collisions occur 45
The Fermi National Accelerator Laboratory at Batavia, Illinois 46
Gordon P. Eaton, head of Hawaiian Volcano Observatory, on rim of Kilauea crater 47
C. Roger Lynds using a powerful telescope at Kilt Peak, Arizona 48
Alec Feiner at his desk at Bell Telephone Laboratories 78
Antennae of a male moth 89
Dr. David Karnosky at the Carey Arboretum in Millbrook, N.Y. 89
Richard W. Robinson with experimental tomato plants 90
Dr. John C. Fiddes of the Laboratory of Molecular Biology in Cambridge, England 91
Professor Thomas R. Odhiambo at insect physiology center at Nairobi 91
Dr. A. Stanley Rand and Stella Guerrero examining a small lizard on Barro Colorado Island, Canal Zone 92
Dr. Ralph A. Nelson taking blood from a bear 92

William Nunn Lipscomb, Jr., with molecular models in his laboratory 93
David M. Phillips studying sperm with an electron microscope 94
Dr. F. Herbert Bormann counting blossoms on a young pin cherry tree 169
Dr. William B. Jackson studying dead rats at Enjebi 170
Dr. Ruth Patrick displays creature netted in a Pennsylvania creek 171
Dr. L. David Mech traces the migratory pattern of a lone wolf 172
Dr. Eville Gorham trekking through a sedge fen in the Big Bog 172
Dr. Roscoe Brady studies the role of fats in hereditary diseases 209
Dr. Edwin Kilbourne testing swine-influenza virus 210
Doctors Joseph E. McDade and Charles C. Shepard discussing discoveries of the cause of Legionnaires' disease 210
Dr. Stanley Falkow working on test of resistance by organisms to antibiotics 211
Dr. Stanley J. Dudrick gives intravenous nutrition to patient with severe ileitis 212

Introduction

Every few weeks there appears in *The New York Times* an article about a scientist and the work he or she is doing. The scientist can be young or old, famous or obscure, looking into man's past or the universe's future, probing the microscopic reaches of cell or atom or the interconnectedness of life in forest or bog or stream, studying island rats or tracking the cause of a mysterious illness. The scientist may be on the threshold of a major discovery or grappling with a line of investigation destined to be little more than a footnote to the sum of knowledge. But in each case the scientist and his or her work make a story worth telling, for the reason given in the explanatory note that accompanies each article:

> SCIENTISTS AT WORK
>
> This is another in a series of articles, appearing from time to time, describing the creative process of scientific research.

Few people in our society dispute the idea that science is important and that scientists are indispensable to the way we live and the way we will live. But there is less certainty as to what science really is and who scientists are. Abstruse language and white-coated, bubbling-beaker stereotypes cloud our perceptions. This creates the impression too often that science is a thing apart, the other of the "two cultures," and perhaps even beyond ordinary comprehension. Consequently, even though scientists may inspire awe and command respect, they have not usually been thought of as "a good read."

Several hundred people in the United States, who call themselves science writers, are dedicated to the proposition that science is not only important but interesting. *The Times* has the largest and most experienced staff of science reporters of any newspaper in the country. Eight full-time reporters are based in the New York office, but their stories can take them almost anywhere in the country or abroad. In addition, one reporter covers science in Washington, and another reports on technology for the Business/Financial section. No member of the science staff has had less than ten years of reporting experience.

More than any other single paper, *The Times* has distinguished itself over the years in the thoroughness and consistency of its coverage of the facts of science—the announcements of discoveries and inventions; the progress reports on research of immediate or potential importance; the news of expeditions in space, oceanography, or archeology; the implications of science and technology on the environment and such global issues as food, health, population, and resource development. But we, like other newspapers, had never made it a regular practice to describe the process of science. How do scientists get their ideas and test them? What are their motivations and pressures, their frustrations and satisfactions? What is the "feel" of doing science in the latter half of the twentieth century?

The reasons, or excuses, for this neglect were both journalistic and scientific.

Consider, for example, what happens in the customary reporting of science news. The news usually surfaces in the form of an announcement from a university or government agency, a paper delivered to a scientific meeting or published in a scientific journal, an interview or chance conversation with scientists. There is often neither time nor space in the newspaper to develop the step-by-step process by which the research was conceived and executed. The story must be reported and written in a few hours or at most a couple of days. It must be written in about 500 to 1,000 words, many paragraphs of which must be devoted to explanations of terms and concepts and possible implications. If any of the "color" about the scientist and how he did his work is included, this is usually the first to be sacrificed in the editing, the victim of layout restrictions or an editor's limited attention span. (Recent newspaper readership studies by the Newspaper Advertising Bureau challenge the assumption of many editors that science stories have slight reader appeal. In 1971, 47 percent of a wide range of listed newspaper topics had a greater readership than the category of science and invention. In 1977, only 26 percent of the same topics received higher reader-interest ratings than science and invention.)

Scientists themselves contribute to keeping the public in the dark about the ways of scientific research. In the 1977 Phi Beta Kappa lecture to the American Association for the Advancement of Science, June Goodfield, adjunct professor at Rockefeller University, discussed the relationship of science and the public in terms of "the problem of humanity in science." One source of the rather bloodless image of scientists, she said, "arises through the inevitable stance of detached objectivity whereby a scientist must approach the natural world." This was what Marie Curie had in mind when she said, "Science deals with things not people."

The scientific journals are replete with examples of the dry way scientists communicate with each other and the public. "It is difficult to reach out and touch the humanity, or the humaneness, in the people who do science," Dr. Goodfield said, "because science is essentially a communal activity whose results must be expressed in the passive voice, to be understood by anyone throughout geographical space and historical time. The expressions of science come in forms from which all the human content has necessarily been drained. So questions, Who are the people who do science as individual human beings? What is the relationship between them and the scientific ideas they create? How and in what form are individuality and creativity brought to bear and expressed in science?—these are pressing questions which have not received the attention they deserve."

If scientists often fail to tell their own stories in human terms, it is all the more necessary for journalists to make the attempt. This was what we set out to do in the *Scientists at Work* series, which was initiated in 1975. This is a collection of twenty-nine articles from the series. It is hoped that by focusing on one or two scientists in each article, telling through them the story of one particular avenue of research, the nature and excitement and variety of science will become more readily understandable.

The selection of subjects for the series was somewhat random. An effort was made to include people working in the "hot" areas of science, such as genetics, biochemistry, and particle physics. A few of the articles concerned subjects in the news at the time, such as scientists working on the swine flu vaccine and searching for the bacterium causing the so-called Legionnaires' Disease. Others were chosen because their work, though still in process, appeared to be headed toward important results or was at least illustrative of an important and/or interesting problem being addressed in a certain field of science.

All the scientists we approached were quite willing to cooperate. Interviews with them often extended over a period of several days, in the laboratory or out in the field. It meant staying up all night at Kitt Peak Observatory and camping out at an archeological dig in Pennsylvania and on the tiny, uninhabited Pacific island of Enjebi. It meant trekking through the California high desert and a New Hampshire forest and flying low over a frozen Minnesota lake on the trail of timber wolves. Whenever possible, the scientists talked about their work where they worked and while they worked. We listened and learned, and then headed for our typewriters to share what we had learned about a scientist and the work of science.

We feel that the results, published here in only slightly revised form, convey the essence of science as an important and fascinating human adventure.

JOHN NOBLE WILFORD

I

In Search of the Past

ARCHEOLOGY, PALEONTOLOGY, DENDROCHRONOLOGY

Autopsy on an Egyptian mummy under way by members of the Paleopathology Association. At right, Dr. Theodore Reyman uses an electric saw, as Dr. Michael Zimmerman stands by. The scene is a lab at Wayne State University medical school, Detroit.

Left: Dr. Henry N. Michael of the University of Pennsylvania Museum examining the base of a bristlecone pine, a tree that may live 5,000 years, in California's White Mountains.

Below: Ralph L. Holloway, Jr., measuring the brain-shaped cast he made of the inside of the skull of an extinct ancestor of modern man for clues to the evolution of the human brain.

Dr. Richard S. MacNeish and students examining material taken from the dig.

Archeological students on a dig in western Pennsylvania where detailed glimpses of ancient life are being unearthed.

1

AUTOPSY ON A MUMMY

Boyce Rensberger

DETROIT, Michigan—Under hot floodlights and before a closed-circuit television camera, Bill Peck gently lifted a loose end of the mummy's fragile linen bandage and began to unwrap the preserved corpse.

Brown dust, undisturbed for more than 2,000 years and pungent with ancient Egyptian balms, swirled up into the nostrils of a dozen scientists clustered around the smallish human form lying on a wooden table.

Making notes and photographs of the wrapping techniques used by the ancient embalmers, slowly they unwrapped about twenty layers of ribbonlike bandage, some of them still soft and strong, some turned brittle and crumbly.

Then, after nearly six hours, Mr. Peck and the scientists exposed the dried body of a woman, perhaps twenty-five years old, who had died somewhere between 2,000 and 2,300 years ago. Though severely shrunken, many details of the body's surface, such as fingerprints, could be seen.

Thus, in a laboratory of the Wayne State University medical school, Mr. Peck, curator of ancient art at the Detroit Institute of Arts, made the mummy ready for the beginning, the next day, of a full-scale autopsy intended to show not only the woman's cause of death but a wide range of other details about health and disease in ancient times.

Although the autopsies are not intended to produce much of immediate practical value, they have already yielded insights that influence thinking about the origin and nature of several diseases.

The autopsy, on a mummy lent by the Pennsylvania University Museum, was the fifth performed by members of a three-year-old group known as the Paleopathology Association, an organization that exemplifies one facet of science that seldom reaches public attention—research motivated not by grant or commercial interest but by curiosity and fascination.

The group's members include about 250 physicians and scientists in sixteen countries, a few of whom get together once or twice a year to unwrap and cut up a mummy by day and party by night. Although the mummy dissection is a brief event, samples of various organs are sent to many cooperating scientists around the country for further analysis, which can take months and years.

Unlike most scientific work, the mummy autopsies and subsequent analyses are strictly volunteer efforts by unpaid individuals. To participate, some drive across the country with their families and stay in the homes of friends. Back home with, say, a bit of bone or tissue, they do the analyses after-hours on university or hospital equipment that exists primarily for other purposes.

Egyptian mummies have been unwrapped and cut open many times in the past but seldom, if ever, have there been complete autopsies involving a large interdisciplinary team or making as detailed an analysis of the preserved tissue as have the members of the Paleopathology Association.

For example, the mummy teeth are analyzed by a scientist at Brookhaven National Laboratory for evidence of toxic metals that may have polluted the environment long before the industrial age. The fingernails go to someone at the University of Texas for similar analysis. For various other kinds of study the cartilages of the voice box go to the University of California, deposits within the lungs go to the Johns Hopkins medical school, eyes go to the University of Illinois, and bone samples go to the University of Colorado. If any plant remains are found, they go to a researcher at Harvard.

"We have all kinds of people interested in this. There's no money in it for anybody. Everybody does it because it's exciting," said Dr. Aidan Cockburn, a sixty-three-year-old British-born physician who, with his wife Eve, founded the Paleopathology Association.

Dr. Cockburn (pronounced Coburn) has devoted most of a professional lifetime to his theory that man evolved in tandem with his parasites and germs. He believes that modern man and his infectious organisms have, over millions of years, shaped each other's present nature and that a study of ancient man and his diseases will throw light on the mysteries of modern disease.

Dr. Cockburn paced about on the fringe of a dozen or so people clustered around the mummy beginning the autopsy as photographers and cameramen shifted back and forth documenting the proceedings from every angle for the scientists.

Dr. George Lynn, a Detroit specialist in diseases of the bones of the inner ear, began the autopsy by cutting off the top of the mummy's head with a small, buzzing saw. As the blade vibrated into the brownish skin over the bone, white bone dust sifted down. In minutes the skull was open and a dozen living heads squeezed together to peer into the empty brain case.

As was found in earlier autopsies, a hole had been punched through the brain case by forcing a tool up the left nostril and the brain removed through the hole.

To remove the ear bones, Dr. Lynn fitted a tubular blade to the saw and bored neat cylinders of bone containing the ears. Back at his laboratory he would cut open the bone to look for signs of an ear disease whose cause is still not known.

The skull, which had become detached from the rest of the body, was handed to Dr. Michael Finnegan, a physical anthropologist from Kansas State University, who sat at a table taking measurements and examining what appeared to be a full set of teeth, heavily worn from a gritty diet but otherwise in good condition.

Back at the autopsy table Dr. Theodore A. Reyman, a Detroit pathologist, supervised the opening of the chest and abdomen with Dr. Michael Zimmerman, a University of Pennsylvania pathologist, assisting.

Although detailed findings would await later study, two questions were at least partially answered in the autopsy itself. One that every mummy poses is how the embalming process was carried out. A second question, which took most of a day to resolve despite the assembled anatomical expertise, was the sex of the mummy.

Although x-rays taken before the unwrapping had suggested the mummy was female, diagnosis on bones alone is seldom certain. After unwrapping, the mummy's fingernails and toenails could be seen to have been painted red—the traditional color for male mummies, yellow being used for females.

"What's this about red fingernails," Dr. Cockburn bellowed as he elbowed his way closer to the mummy. "It shouldn't have red fingernails."

"Maybe this is a small boy," offered Dr. Robin Barraco, a Wayne State University biochemist who is hoping to reconstitute some of the mummy's tissues and extract proteins that can be analyzed. He has already done this with earlier specimens in an attempt to find antibodies that can be reactivated and tested to see what diseases the person may have had.

The flesh in the pubic region was not well preserved, but crumbly and deformed. At one point a Toronto pediatrician was convinced he could make out the shape of a penis while, at the same time, someone else saw what he thought were flattened breasts.

It was not until the body was opened up and a uterus found that everybody agreed the mummy had been a woman.

Dr. Reyman made the initial incision, cutting through the chest wall with the electric saw. The hardened flesh broke away in chunks, revealing blackened and shriveled objects and pieces of cloth. The presence of the cloth was initially a mystery because no embalmer's incision could be found.

"Hey," said one scientist, "is that the aorta?"

"I don't know. Looks like it could be," said another, lifting out an irregular shape. "Mark it down 'possible aorta.' "

"Could somebody please bring me some more sample bottles?"

"How did this get in here," someone asked, holding up a wad of cloth.

"You know what I think? They went in through the rectum."

"Could be."

Bit by bit more pieces of the chest and abdominal wall were sawed away and organs removed by hand. Although no one could be sure at the moment, it appeared as if the mummy had a heart, two lungs, a liver, two kidneys, a bladder, and a uterus.

When the lowest portion of the abdomen had been cleaned out, it appeared the body had, indeed, been embalmed through the rectum. This is a technique described by Herodotus, writing in the fifth century B.C., as being used on the less wealthy classes of people who could not afford more elaborate mummifications.

Although no evidence of disease was immediately evident in the latest mummy, earlier autopsies have uncovered infestations of tapeworms, roundworms, and schistosomiasis.

One mummy, studied in Toronto by the same group, appeared to have a ruptured spleen with remnants of the fatal hemmorhage still visible. That mummy, a fourteen-year-old boy, showed evidence that his bone growth had stopped temporarily three times in the last fifteen months of life. These signs, visible as lines in thin sections of bone viewed under a microscope, are considered the result of major infections.

Most of the mummies examined microscopically thus far have had substantial carbon deposits in the lungs, thought to be a consequence of smoky cooking fires, and plaques on the artery walls that doctors usually associate with heart disease.

Although autopsies on other mummies will undoubtedly yield more insights, the spare-time scientists of the Paleopathology Association have already found evidence that two of the conditions usually considered consequences of modern society—air-pollutant contamination of the lungs, and heart disease—may have been with mankind for at least 2,000 years.

2

TIME AND THE BRISTLECONE PINE

John Noble Wilford

BISHOP, California—"You realize, of course, how much of a needle-in-a-haystack job this is," Henry N. Michael remarked, shifting into a lower gear and steering the Toyota Land Cruiser across the dusty alluvial plain toward Silver Canyon.

Dr. Michael, a sixty-two-year-old anthropologist, geographer, and dendrochronologist, was looking in the high desert at the foot of California's White Mountains for old buried logs of the bristlecone pine. The older, the better—preferably 8,000 to 9,500 years old.

"I know there's wood somewhere in here," said Dr. Michael, though his weeks of searching—by radar probe, digging, and educated guessing—had not been overly encouraging. "It's a matter of time and luck."

His was the sort of quest, a lone scientist out in the desert looking for buried logs, likely to make laymen shake their heads in despair of ever understanding the ways of science. Yet it was the quintessence of scientific curiosity, the drive to find a seemingly humble piece of evidence and thereby expand the greater sum of knowledge.

For if the logs that Dr. Michael sought could be found, their growth rings should give archeologists a key to checking and correcting the radiocarbon dates of many prehistoric artifacts. This could result in important revisions of prehistoric chronologies about how some civilizations developed.

Archeologists used to trust implicitly the chronologies derived by the radiocarbon method of dating. The method is based on the principle that all living things absorb radioactive carbon 14, which is formed in the upper atmosphere by the reaction of ordinary nitrogen with neutrons produced by cosmic rays. When the organism dies, the absorption halts and the radiocarbon in its system begins to decay at a slow, constant rate.

Geiger-counter measurements of the decreasing ratio of radioactive carbon to normal carbon give the time since the material lived. But errors in these dates were recently detected in archeological objects (linen, human bones, wooden coffins) from early Egyptian dynasties whose ages had already been established with the aid of kings' lists and astronomical reckonings.

One basic assumption of radiocarbon dating turned out to be invalid: the biospheric inventory of carbon 14 has not remained constant through time. Materials that absorbed greater amounts of carbon 14 would appear to be younger than they really were. This uncertainty in radiocarbon dating sent scientists out on the trail of the bristlecone pine.

Tree-ring dating—dendrochronology—is considered to be nature's most precise chronological technique. And no tree is more ancient than the bristlecone pine. It is the world's oldest known living organism, with life spans that may reach 5,000 years, or nearly 2,000 years beyond the sequoia's longevity.

These pines are rare and grow high on the arid mountaintops in the western United States, anchored in dolomitic soil where little else can grow, gnarled by time's vicissitudes, polished to a fine patina by wind and icy snow—stark and beautiful survivors of a harsh environment.

They survive because of, not in spite of, adversity. With

little moisture, they grow ever so slowly, adding less than an inch of wood to their diameter every century. As protection against the elements, they develop dense, highly resinous wood that is strongly resistant to the infestations that plague most trees. Their growth is sparse, which reduces the threat of spreading forest fires. And their very isolation, at elevations of 9,000 to 11,000 feet, spares them from the woodsman's ax.

In the Ancient Bristlecone Pine Forest, a 28,000-acre federal preserve in the White Mountains near the California-Nevada border, one tree, named Methuselah, is 4,600 years old and still growing. Its exact location is kept secret as a precaution against vandalism.

But visitors are permitted to see Pine Alpha, a 4,300-year-old tree so named because it was the first to be dated older than 4,000 years. A sign at its base reads:

"Pine Alpha had been growing on this slope for 500 years when early man developed the spoked wheel; it was 1,500 years old when Moses led his people out of Egypt, and 2,800 years when Rome fell."

This forest is the outdoor laboratory of the Known Age Project, funded by the National Science Foundation and involving scientists from the University of Arizona's Laboratory of Tree Ring Research, the University of California at San Diego, and the University of Pennsylvania physics department and its Museum's Applied Science Center of Archeology. Dr. Michael is a research associate at the Pennsylvania center and a professor of geography at Temple University.

The purpose of the project is to compile a master chronology of tree ring dates stretching as far back as possible.

Year after year, separately and sometimes together, Dr. Michael and Dr. C. Wesley Ferguson of the University of Arizona have combed the slopes of the White Mountains looking for ever-older bristlecone pines, living or dead, standing or fallen.

On each expedition the scientists took pencil-thin core samples from living pines and sawed off cross sections of fallen trees, many of which had remained in solid condition several millenniums after their deaths. The samples were sent to the University of Arizona for microscopic analysis of the rings.

"It is more than just counting rings," Dr. Michael explained. "It's a long, tedious, imaginative process of looking for signatures."

In trees such as the bristlecone pine, a distinct growth ring is produced each year. Thin-walled cells grow during the spring and early summer, and these contrast with the smaller, thick-walled cells of the late growing season.

Dendrochronologists look for the usually clear line between the late wood of one season and the next spring's growth—and then count the lines inward from the outermost, or bark, ring to the center, assigning an absolute calendar date to each succeeding ring.

Because year-to-year variations in climate are nearly the same, a tree in any given century will produce several distinctive patterns of narrow and thick rings. These are the tree's signatures, and they enable scientists to extend their master chronology beyond the lifetime of any living bristlecone pine.

They do this by cross-dating from both living and dead trees. A piece of dead wood will exhibit signature patterns that partially overlap ring patterns in living trees whose ages are known. If part of the dead tree is older than any part of the living tree, then the chronology can be extended farther back in time.

Dr. Ferguson and his group have thus reached back 7,400 years from the present and have one specimen that is 8,100 years old.

After the ring analysis, parts of the samples undergo radiocarbon testing, primarily at the universities of Pennsylvania and Arizona. The older the sample, it has turned out, the greater generally the divergence between the tree-ring age and the radiocarbon date. The maximum divergence is at present at about 5000 B.C., where the radiocarbon method gives a date at least 750 years younger than the material really is.

From such calibrations scientists have prepared tables to show the correction factor that must be applied to any radiocarbon date. One of the first tables, compiled by Dr. H. E. Suess of the University of California at San Diego, was responsible for revising certain European prehistoric dates.

As a result, archeologists say, it now appears that the "barbarians" of western Europe developed tools and built megalithic tombs and monuments independent of influence from the supposedly more advanced civilizations of the eastern Mediterranean.

But does the dating divergence keep widening the farther back one goes? This question brought Dr. Michael back this year to the alluvial plains below the White Mountains.

"Since Ferguson and I got into this," Dr. Michael was saying on the way to a digging site, "the chronology has been pushed from 700 B.C. to about 5400 B.C. That's progress of 400 years a year—only it doesn't always work that way. For the last three years we've been stymied. We can't find anything any older up on those slopes. I guess that's why I'm looking for some older wood in unconventional places."

Dr. Michael spoke with a vague accent. Though born in Pittsburgh, he spent his teenage years in Czechoslovakia between the world wars. A job cataloguing shards at the University of Pennsylvania Museum drew him into archeology and anthropology. His Ph.D. dissertation was on the Neolithic Age in eastern Siberia, and he has continued research in the subject with field trips to the Arctic and as an editor and translator of Russian publications on Siberian archeology and ethnography.

But now he was in bristlecone pine country—a scientist on location, as it were. His unlined face was tanned from the sun and wind, his dark brown hair slightly bronzed. A cooler with the day's provisions (apples, water, and bitter lemon) rattled in the back of the Land Cruiser.

Pointing to the narrow entrance to Silver Canyon, Dr. Michael explained the "mental process" underlying his search strategy.

"According to geological evidence," he said, "about every 400 or 500 years there is an exceptionally strong thunderstorm here that washes the big boulders and fallen trees down these slopes into the canyons and eventually out onto these alluvial fans. We're trying to find where some of those logs may still be buried."

A few days earlier, a team from the Stanford Research Institute had conducted a reconnaissance using a newly developed system that penetrated the dry soil with radar signals. Tests at the sites of Indian ruins in New Mexico indicated that the return radar signals could, if carefully analyzed, locate buried boulders and logs.

"The radar hasn't worked here so far," Dr. Michael said, stopping at the site of one disappointing dig. "I think I'm looking too low on this fan."

But on several "blind digs"—decided on by hunch, not radar—the scientist found some encouragement. He uncovered a logjam buried in the upper reaches of the alluvial fan at nearby Millner Canyon. But the logs were not nearly old enough, only 2,200 years old. On the upper slopes of Silver Canyon Dr. Michael dug up a bristlecone pine cone that lived 5,000 years ago.

Dr. Michael also sought to enlist the help of miners who

worked the small silver and platinum claims up in the canyons. But their response was usually a hard-eyed suspicion. What did this scientist really want?

A similar request at the gravel quarry down near Owens River, however, brought results. One of the workers rooted out a log and gave it to Dr. Michael. Radiocarbon dating disclosed it to be 13,000 years old. But it was rosewood, not bristlecone pine. Though useless for extending the master chronology, the piece of rosewood renewed Dr. Michael's hopes for eventual success.

"We have found wood buried in the alluvial fans, but it's not as old as we want," he said. "And we have the rosewood from the quarry. We'll try again next year."

On the drive into Silver Canyon Dr. Michael surveyed a number of promising sites for next year's diggings. He would search in the moist soil near where the creek runs out of the canyon. And he would give the radar one more chance.

Dr. Michael is a patient man. He has to be in his work of finding buried logs, counting tree rings, and dating millenniums.

"Eventually, dendrochronology will be carried to 10,000 and 11,000 years," Dr. Michael concluded. "I think it can be done right here in the White Mountains."

3

A NEW LOOK AT OLD SKULLS

Boyce Rensberger

NEW YORK—Ralph Holloway cradled the ancient skull in his hands, holding it upside down as he peered through a hole into the empty cranial vault. Somewhere between two and three million years ago it housed the living brain of an early form of human being.

Dr. Holloway, who was working as a visiting scientist at the National Museums of Kenya in Nairobi, was examining the famous "1470" skull that Richard Leakey found in the wilds of northern Kenya. He turned the fossilized, brownish bone to catch the light against the undulations on the inside surface.

For each depression on the inner wall of the skull, Dr. Holloway knew, there was a corresponding bulge on the brain. Some of the bumps and grooves on the brain are known to mark regions specializing in specific mental and behavioral functions.

Dr. Holloway was hoping to learn from this long-buried skull something of how its brain was organized. From that and from studies of other fossil skulls of extinct forerunners of modern man, he is attempting to look beyond the bony remains to gain some hint not just of what our ancestors looked like but of how they behaved and even how they thought.

Ralph L. Holloway, Jr., a soft-spoken, full-bearded professor of anthropology at Columbia University, specializes in a form of research so new—and to some so seemingly unprofitable—that he may be the only scientist doing it. He makes rubber latex casts of the insides of skulls of early man and, from the brain-shaped casts, tries to deduce something about the evolution of the human brain.

Dr. Holloway's work has sometimes been compared with phrenology, the long discredited pseudoscience linking the outer shape of the skull to individual personality. In fact, his work represents a very recent application of modern neurological concepts, topographic analysis, and evolutionary theory to the newly discovered wealth of fossil skulls from East Africa.

Much of the work is done on the Columbia campus in a cluttered laboratory lined with books, casts, and leering skulls of various monkeys, apes, and early human beings. A gorilla skeleton stares from its glass case across a table stacked with computer printouts at a human skeleton near the opposite wall.

About once a year Dr. Holloway goes to Africa and sometimes to Asia to examine new finds and to join expeditions, trudging along dusty hillsides to look for fossils.

The most remarkable thing that Dr. Holloway found on a cast he made inside the 1470 skull was evidence that the individual who possessed it more than two million years ago (the exact age is in doubt) was capable of some kind of human language.

The evidence is a bulge known as Broca's area that is lacking in apes but present in man and is considered a speech center. This evidence is far from conclusive, but the implication is revolutionary. Until this discovery the oldest accepted evidence of linguistic ability in man dated back only a few tens of thousands of years.

Dr. Holloway has made scores of other casts of extant ape

species, extinct forerunners of man and modern human beings. His analysis and interpretations of the endocranial casts, or "endocasts," and his expertise in estimating the cranial capacity of fossil skulls, many of which are incomplete, have made him a highly regarded authority on the evolution of the human brain. Much of his work is sponsored by the National Science Foundation.

"My endocasts are never going to prove conclusively that a creature had language ability," he said between sips of lapsang souchong, one of a dozen exotic teas he stocks and brews in his lab. "What I'm trying to do is to see whether we can't look at these casts and see whether the relationships between various parts of the brain have changed with evolution. If we could see certain changes, we might be able to correlate them with evidence of behavioral changes from the archeological record. Then we might come to understand our past a little better."

Before the forty-one-year-old scientist began his work fourteen years ago, examining the internal structure of rat brains at the University of California at Berkeley, about the only thing known about the evolution of the human brain was that it grew larger than those of apes. At some arbitrary point, it was held, the brain became big enough safely to be regarded as human.

Dr. Holloway holds that size alone is not nearly as important as internal organization. There are key differences between the way an ape's brain is put together and a human's, and hazy but useful markers of some of these differences can be seen on the casts.

For example, the boundary between the occipital, or visual, lobe, in the back of the brain, and the parietal lobe, just forward of the occipital, where much distinctively human thinking is done, is a groove known as the lunate sulcus. This feature can be seen on endocasts and its position indicates whether a relatively larger share of brain volume has been devoted to one part of the brain or the other.

In monkeys and apes, the lunate sulcus is positioned toward the front of the brain, indicating a largish visual lobe and a smallish thinking lobe. In modern human brains, the boundary is toward the back of the brain, indicating a reduction in the size of the visual lobe and an enlargement in the thinking regions.

Endocasts of extinct predecessors of modern man such as *Australopithecus* and the small-brained *Homos* show the lunate sulcus in the human position even though the total brain size may be barely larger than that of an ape.

Dr. Holloway speculates that this reorganization of the brain was related to the appearance of a wholly new form of behavior including such characteristically human attributes as language, tool making, and a closely knit social structure centering on a home base. He suggests further that once this broadly useful evolutionary step had been taken, it granted access to a variety of ecological niches that no previous animals had been able to exploit. Populations of those early humans adapted to different niches, an evolutionary phenomenon known as adaptive radiation, and went on to become the various contemporaneous forms of early man that recent fossil discoveries indicate were living between one and three million years ago. All but one of the lineages eventually died out.

The starting point for these theories is the endocast which, to the untrained eye, may look like a misshapen blob of rubber. In fact, close inspection often reveals not only the broad contours of the brain but details such as the paths of tiny blood vessels that once lay between the brain and the skull.

"Each skull that I do has to be approached on its own terms," Dr. Holloway said, referring to the fact that most fossil skulls are actually many fragments of shattered bone that have been glued together again. The reconstructions may be fragile and missing several pieces.

After the gaps have been filled with clay to seal the brain case, Dr. Holloway pours a puddle of liquid rubber latex through the foramen magnum, the opening for the spinal cord, into the cranium. He sloshes it about to coat the entire inner surface, pours out the excess, and lets the rubber dry for an hour or more. Six or seven more layers must be poured inside the skull to build up a sturdy thickness. Skull and rubber are then cured in a 140-degree oven for four hours to set the shape.

If the skull is whole and cannot be disassembled into pieces, Dr. Holloway collapses the dried latex and pulls it out through the foramen magnum, and its "elastic memory" springs it back to the proper shape. The rubber is then filled with plaster of Paris to make the cast rigid.

In cases where the skull can be taken apart and later reglued, Dr. Holloway prefers to put in the plaster before removing the rubber from the skull.

This was the technique used on the 1470 skull. Working in a Nairobi laboratory of the National Museums of Kenya, Dr. Holloway allowed the plaster to harden inside the skull and then carefully brushed acetone along certain joints to dissolve the glue. He had planned to separate the cranium into four pieces, remove the cast, and reglue the fossil.

"It fell apart beautifully," he recalled. "The first thing I did was look for Broca's area. I had suspected it from earlier examination of the cranium. I was immediately impressed with it."

Dr. Holloway then measured the cast to check for accuracy against the inner dimensions of the skull, and assured himself that it was virtually distortion-free. The whole job took a week and a half.

While some features of the cast, like the Broca's area, are immediately apparent to the trained eye, Dr. Holloway suspects that many others that are not may be revealed through more sophisticated topographic analysis of the cast's shape. In his Columbia laboratory he uses two mechanical devices to measure the subtle contours of each cast. One device, designed and constructed by Dr. Holloway of precision machined aluminum and brass (his early training was as a metallurgical engineer), is called a three-dimensional coordinate caliper system.

It is used to detect lopsidedness, or asymmetries, in the brain. The brains of lower animals are generally symmetrical, the right hemisphere being a mirror image of the left. It is assumed that their functions are mirrored also. But, as evolution approaches man, asymmetries begin to appear. In humans these are known to be related to various specialized mental abilities. It is assumed that if an extinct ancestor of man had similar asymmetries, they were related to the emergence of these mental abilities.

The other device, designed by Alan Walker, a Harvard University anatomist, and built of Plexiglas, is called a polar coordinate stereoplotter. The device makes it possible to measure the distance from the center of the cast to any point on the surface. Dr. Holloway, with the help of his graduate students, is currently taking such measurements at about 300 points evenly spaced over the surface of each cast.

From these numbers, he plans to draw contour maps of the casts and, using such mathematical techniques as trend surface analysis and multivariate analysis, compare the contours to see whether those from one species of early man fit a pattern distinct from that of another species.

Preliminary analysis of some of the contour data indicates that the casts do fall into species groups on mathematical terms alone.

"I realize that 300 data points on a cast may be more than we need. I have to admit this may be simple mathematical overkill," Dr. Holloway remarked. "But once the basic data are established and if I know what I am doing, we may have a very powerful analytic tool."

This summer he plans to go to Nairobi for a few weeks to examine, among other things, a newly discovered skull of *Homo erectus* that is about a million years older than previously known fossils of this human ancestor. He plans to make a cast of the brain case.

This new find, like so many in recent years, has strengthened the view that man's lineage is much older than had been supposed. Dr. Holloway's discovery of essentially human patterns in the very early, small brains of various kinds of early man has added a significant new dimension to this view, strongly suggesting that man's most distinctive attribute did not suddenly appear recently as the capstone of human evolution.

Rather, Dr. Holloway's work is showing, an essentially human brain, qualitatively different from that of the apes, was in existence at the earliest known stages of man's emergence, before the human body looked very different from that of the apes.

4

A REVISED LOOK AT MAN'S ENTRY INTO THE NEW WORLD

Boyce Rensberger

La Jolla, California—On a grassy bluff 325 feet above the Pacific Ocean beach, a team of scientists and students is digging up evidence of what life was like for a people who are believed by some to have lived here 40,000 to 50,000 years ago. Their archeological excavations are part of a revolution that has been underway for several years in the understanding of how and when man first entered the New World.

Until recently, it had been a virtual dogma among American archeologists that the peopling of North and South America did not begin until about 12,000 or, at the most, 20,000 years ago when bands of big-game hunters crossed the Bering land bridge from Siberia into Alaska.

In recent years, however, new archeological discoveries and a new way to date fossil bones have yielded substantial evidence that man's antiquity in the Americas may reach back to 70,000 or more years ago. These new lines of evidence, however, are not accepted by all authorities in the field and much controversy remains about their validity.

One implication of the new interpretations is that modern man reached the New World long before he reached Europe. In fact, as matters stand now, the oldest known skeletal examples of modern man anywhere in the world, according to a controversial new dating method, come from California. It is believed that modern man originated somewhere in the Middle East, and still older bones may someday be found there.

Ironically, some of the best new clues have come from bones that were discovered half a century ago and have been lying in museums ever since. Although various anthropologists over the years had suggested that man entered the New World more than 12,000 years ago, their evidence failed to persuade most of their colleagues.

But in 1974 a young La Jolla chemist named Jeffrey Bada created an uproar in anthropology by applying his own new method of dating to some of the museum bones. One, a distinctly modern-looking skull from an oceanside cliff in nearby Del Mar, he found to be a wholly unexpected 48,000 years old. The oldest known examples of modern man, sometimes called Cro-Magnon Man, from anywhere else are about 40,000 years old. Another set of human bones from La Jolla turned out to match in age the oldest Cro-Magnon remains from the Old World.

The place in Del Mar from which the older bones came has since been destroyed, but the La Jolla site remains at least partly intact.

"When I was asked if I wanted to dig the La Jolla site, I thought about it for maybe a second and a half," said Dr. Jason W. Smith, a young archeologist from California State University at Northridge. "I think we can lay the 12,000-year myth to rest," he added. "I don't care how old the La Jolla site is, as long as it's more than 12,000 years. Once that's established then we can get on with the much more interesting questions about how they lived and who they were."

Dr. Smith, a codirector of the excavation, is the author of a new text called *Foundations of Archeology.*

Although a number of sites believed to be older than 12,000 years are known, they either contain human bones without stone tools or they yield crude artifacts without human remains to certify them as man-made. Dr. Smith is hoping to find both together in unambiguous stratigraphic association.

For all the dust rolling out of the straight-walled, square pits as Dr. Smith's students patiently chip and brush in the hot sun, the location might as well be a remote wasteland. In fact, it is a wealthy residential suburb of La Jolla. More specifically, it is the backyard of the chancellor of the University of California at San Diego, on a narrow strip of land between his swimming pool and the edge of a bluff overlooking the ocean.

It was on this site, according to fifty-year-old field notes, that the 40,000-year-old bones were found. Before beginning to dig, Dr. Smith and the project's other codirector, Dr. Gail Kennedy of the University of California at Los Angeles, who is an expert on fossil bones, walked over the ground and down the crumbling hillside to see whether erosion had exposed any more fossils or artifacts.

It had. They found a number of chipped stones, some crudely flaked and evident only to a trained eye as primitive tools, and a few beautifully sculptured into unmistakeable points and knives.

Because they had eroded out of the level at which they were dropped by their makers or users, there was no way to know how old they were. Stone tools can be dated reliably only when they are found in clear association with other things that are directly dateable, such as bones. Such associations are found by digging very carefully in undisturbed earth and recording the exact location of each object found.

"Where you find one artifact, there's usually more," said Dr. Smith. "We opened up a pit and luck was with us."

With stakes and string, Dr. Smith and his students marked off the ground into a checkerboard of squares, each two meters on a side, and began digging in nonadjacent squares. The idea is both to sample the area widely without having to dig up everything, and to leave part of the area intact for future scientists who, as Dr. Smith put it, "know more than we do."

The digging is done in stages, proceeding down one ten-centimeter layer at a time. Using trowels, ice picks, and delicately manipulated shovels, the dirt is loosened. When a large object is noticed, the digging shifts to smaller tools such as dental picks and brushes to remove the dirt without disturbing the object. All the loosened dirt is shoveled into a sieve, or screen, next to the pit, where another person sifts the dirt from any small bones, teeth, or stone flakes that might have been missed in digging.

About two feet down in the La Jolla pits, excavators encountered a number of rocks—some rounded cobbles and some chipped tools—and bits of bone. It was what archeologists call a "living floor"—an old land surface on which people had once lived long enough to leave signs of their presence.

Some of the bone was given to Dr. Bada for dating with his new technique, called amino acid racemization. Without this method the living floor could not have been dated, for the only other method, carbon 14, consumes more bone than was available from the dig. Dr. Bada's method works on a fraction of an ounce. Until now much archeological evidence has remained undated because the researcher was unwilling to destroy it in the dating process. Also, the carbon 14, or radiocarbon, method is useless for materials older than about 40,000 years, whereas racemization can be applied to objects up to about a million years old.

The new method is based on the fact that amino acids, the

building blocks of proteins, can exist in two chemically identical forms that, in structure, are mirror images of one another. In analytical devices, one form will deflect polarized light to the left while the other deflects it to the right.

In living bone protein, all the amino acids are in the so-called left-handed form, but in death the amino acids, which often do not deteriorate, begin "flipping" into the right-handed form. This phenomenon, called racemization, occurs at a relatively consistent rate over thousands of years. The rate is, however, influenced by the heat and water to which a bone specimen has been subjected over the years, and this variability causes much of the doubt about its accuracy. A measurement of the relative proportions of left- and right-handed amino acids will, however, give some indication of how long the process has been going on.

Dr. Bada took the bone to his laboratory at the Scripps Institution of Oceanography, about a mile down the coast in La Jolla. There, in a jungle of glass tubes, flasks of boiling acid, condensers, filters, and other equipment, Dr. Bada cleaned the specimen and dissolved it in acid to liberate the amino acids. Then he put the dissolved amino acids into a $20,000 machine called an amino acid analyzer. This commercially available device, an automated laboratory the size of a telephone booth, prints out a graph showing the relative amounts of the two forms of each amino acid.

The racemization process determined that the bones were between 7,000 and 8,000 years old.

Back at the dig, the students had exposed the entire living floor, drawn the positions of the artifacts on a diagram of the pit, photographed the floor, and removed the stones to continue digging.

Several days later, at a depth of almost four feet, the excavators uncovered more bone. After carefully scratching away the encrusted, sandy soil, Dr. Kennedy determined it to be a human burial. She found the crumbling leg bones of a person who, she deduced from the position, had been buried on his back with the knees drawn up against the chest.

Dr. Bada took a small piece for dating and found that the person had died between 17,000 and 20,000 years ago. In a nearby pit, at about the same depth, other students found a living floor with some two dozen smooth and chipped stones. One smooth stone was heavily worn, suggesting that it had been rubbed against another flat stone in a manner still used today to grind grain for cooking.

Such stones are most typical of an agricultural people, and agriculture is not known from anywhere in the world of 20,000 years ago. Dr. Smith believes that the Californians of that time were not cultivating grains but, more likely, gathering wild seeds.

Instead of trying to lift the fragile leg bones out, excavators cut away the surrounding dirt and removed the entire hardened block of soil in which the bones were still largely embedded. Dr. Kennedy took the whole block to a laboratory at Scripps where she can take the time to remove the bones more carefully while the digging in the pit goes on. The dig has not found anything to match the 40,000-year age of the bones discovered there fifty years ago.

Also taking part in the dig is Dr. Richard S. MacNeish, an archeologist whose excavations in the Pikimachay Cave in Peru have established the presence of man at several stages ranging in age from 9,000 to 20,000 years ago. Dr. MacNeish, director of the Peabody Foundation for Archeology, is a leading exponent of the view that man entered the New World about 70,000 years ago. That is the most recent time when sea levels were low enough to expose the Bering land bridge so that people could have crossed into North America in time to reach southern California by 48,000 years ago (the age of the Del Mar skull).

Although archeologists tend to think of man as entering the New World only when the land bridge was available (mainly during Ice Age maximums), some argue that even today it is possible to walk from Asia to America in severe winters when the ice freezes over the Bering Straits. It is known that entry during an Ice Age would not have been hampered by glaciers, for there remained an ice-free corridor reaching from Alaska southward through central Canada and into the western United States.

Dr. MacNeish and a growing number of his colleagues believe that when man arrived, he brought with him no tools better than stones crudely shaped into irregular choppers and flakes. Until recently most archeologists believed that the earliest stone tools of the New World were the beautifully shaped and deadly Clovis and Folsom spear points, named for their places of first discovery in New Mexico.

Dr. Paul Martin of the University of Arizona has gained considerable attention in recent years with his theory that the first Americans, already highly skilled hunters equipped with such weapons as they crossed into North America 12,000 years ago, were responsible for the extinction of mammoths, mastodons, giant ground sloths, and other large mammals that took place about then.

Dr. MacNeish disputes this. He cites the growing body of evidence that man was here long before the great wave of extinctions and that he was using tools much more primitive than the Clovis or Folsom points.

"The available archeological evidence not only fails to support the Martin hypothesis but tends to refute it," Dr. MacNeish concluded after analyzing seventy-five archeological sites in the New World, including twenty-three that have yielded artifacts or skeletons that have been radiocarbon dated at more than 12,000 years of age. The oldest, near Lewisville, Texas, is between 37,000 and 40,000 years old, the limit of the radiocarbon method.

One archeological leader who disputes the older dates for man's entry into the New World is Dr. Robert Heizer, now retired from the University of California at Berkeley.

"Man may have come into the New World 40,000 years ago or longer. There's no reason he couldn't have," Dr. Heizer said in an interview. "But there's just no good evidence for it and until we have that evidence, I don't think we should say it happened then." Dr. Heizer believes the dates produced by Dr. Bada's racemization method are suspect. The thing that is wrong, he says, is that "they're too old."

Dr. Bada has heard the same reaction from other anthropologists. "When we first published this in *Science,* I was very naive about the emotional reaction from anthropologists," Dr. Bada said. He added that physical scientists who reviewed his methods were unanimous in their praise but that anthropologists who believed man was a latecomer to the New World asserted that the method must be faulty.

Dr. Bada has used his method on archeological sites that have also been dated by the long-accepted carbon 14 method and found consistent agreement. His dating of bones from African and European sites beyond the carbon 14 limit has agreed well with dates derived by other methods. "I don't see why the technique should work everywhere else in the world but not here," Dr. Bada contends.

"If Jeff's dates hold," said Dr. Kennedy, referring to Jeffrey Bada, "then you have something resembling Cro-Magnon Man in the New World before you have him in western Europe."

Although it may take anthropology some years to completely accept the idea of a human arrival in North America some 70,000 years ago, at least one archeologist thinks even that revision in orthodoxy will not be enough.

Dr. George Carter, who worked as an archeologist in San

Diego decades ago and is now a professor at Texas A&M University, is convinced he has found evidence that man was in the San Diego area at least 100,000 years ago, and has argued his case for many years. He also says there is some evidence from Mexico of man living there 250,000 years ago.

Dr. Carter's theory has long been discounted by most archeologists. The chipped stones that he says are man-made are called "Cartifacts" by others presumed to be the result of natural processes. Dr. Carter, now sixty-four years old and relishing his position as a maverick, is spending the summer in La Jolla, visiting the digs frequently and offering all kinds of advice. Although his work is not even mentioned in Dr. Smith's textbook, no one minds his presence, for it was Dr. Carter who started the current revolution by asking Dr. Bada to date the Del Mar skull that everyone else had forgotten about.

5

CLUES TO THE LIFE OF STONE-AGE MAN

Boyce Rensberger

AVELLA, Pennsylvania—Chill autumn winds howled across a massive rock overhang sheltering the ledge where a little band of Stone-Age hunters huddled around a campfire. The men, bundled in animal skins, used sharp flint knives to cut up a deer's leg for roasting. The women, similarly clad, sorted baskets of nuts.

Such a scene, or one very much like it, took place 16,000 years ago on a steep hillside just outside this far-western Pennsylvania town. It was the beginning of the end of the last Ice Age and, geological clues indicate, the edge of the glacial ice sheet was barely fifty miles to the north as descendants of the first people in North America ranged the forests of Pennsylvania hunting deer and elk and gathering walnuts, hickory nuts, and hackberries.

Evidence of such prehistoric scenes, apparently reenacted many times by aboriginal Americans for thousands of years, has gradually been uncovered in one of the most remarkable archeological digs ever undertaken in North America. The digs are on the oldest known site of human occupation in the New World.

The excavations by a team of scientists and students led by James M. Adovasio were carried out every summer from 1972 through 1977 on the floor of what is known as the Meadowcroft Rockshelter. The site is just a few yards up a hillside from Cross Creek, one of only two easy-to-travel valley bottom routes ancient peoples would have used to escape the interior of Pennsylvania for milder winters to the south.

Archeologists have determined that the rock shelter was a major stopover for hunting and gathering bands traveling through the region. Among the traces they left behind are charcoal hearths, baskets woven from bark strips, stone tools, and the knife-scarred bones of their prey.

Isolated bits of evidence of still earlier human presence in other parts of the continent, reaching back to 40,000 years ago or more, are accepted by some but not all experts. In no other location, however, is there layer upon layer of solidly dated evidence that people were living in North America from as far back as 16,000 years ago.

There is also a fragment of evidence—a strip of bark that appears to have been cut as if for a basket—from the deepest layer of Meadowcroft suggesting that people were there as early as 19,000 years ago. Although such evidence would be readily accepted by many other archeologists, the scientists directing the Meadowcroft dig have chosen not to put too much weight on it until there is corroboration.

"I don't see any need to lean too heavily on that oldest material at this point," said Dr. Adovasio, the thirty-four-year-old archeologist who has led the Meadowcroft excavation since it began. "The other stuff we have is so securely dated that I'm quite content to make a case on that for man having been in the New World quite early."

Dr. Adovasio, a muscular man with a close-cropped black beard, is an associate professor of anthropology at the University of Pittsburgh, about thirty miles east of here.

Despite the strong contention by some that the peopling of the New World began 40,000 years ago or more, the evidence for human presence here prior to 12,000 years ago has not constituted an airtight case. Meadowcroft, with its dozens of layers of dateable human traces ranging from at least 16,000 years ago up to A.D. 1265, is rapidly breaking down the 12,000-year barrier.

If humans were in Pennsylvania 16,000 years ago, Dr. Adovasio suggests, they must have crossed into the New World from Asia perhaps 30,000 years ago. One possible link between Meadowcroft and Asia is that the oldest stone tools from the Pennsylvania site resemble early tools found in Siberia.

Less speculative is the evidence of agriculture from Meadowcroft, remains of domesticated corn and beans dating from the first millennium B.C. Although farming is known to have begun much earlier in Middle America, the Meadowcroft finds indicate an earlier penetration of the East than had been documented before.

Five days a week, from 6 A.M. to 6 P.M., and half a day on Saturdays, the rockshelter is busy as Dr. Adovasio and eighteen archeology students dig with trowels and brushes. The main excavation, just over sixty square meters in area, has been roofed and brightly lighted with experimental lamps that bring out subtle color differences in the layers of soil and prehistoric debris that have buried successive "living floors" in the shelter.

While the more typical dig has neat straight-walled square pits with relatively flat bottoms, excavations of rockshelters must work around massive boulders that have fallen from the roof. Such falls distort the deposits below and provide an irregular base for subsequent deposits.

One huge roof fall at Meadowcroft in 165 B.C. apparently quenched a wood fire that was burning in the shelter at the time. If there were people around the fire when it happened, their bodies may be recovered some day when the excavation tunnels under the fallen rock.

Digging through a cubic meter in the shelter with trowels and brushes carefully enough to identify, mark, and record every stratum, artifact, bone, and seed can easily take two persons an entire summer.

"Digging a site destroys it, so we've got to get as much information as we can the first time," Dr. Adovasio said one afternoon as he sat at a table perched precariously on the steep slope just outside the roof over the dig. Nearby half a dozen students sifted buckets of dirt through a sieve, looking for anything that might have been missed by the excavators.

Suddenly Mike Beckes, the crew foreman, came running up to Dr. Adovasio with something in his hand. "Good loot," he shouted. He handed Dr. Adovasio a blackish, disc-shaped piece of flint that had been flaked from both sides to make a sharp edge all the way around.

"Fantastic," Dr. Adovasio exulted. "Oh, this is nice! This is a truly neat thing!"

The object, a discoidal biface in archeological parlance, had come from a stratum just over 12,000 years of age. Dr. Adovasio took the artifact into the dig, still exclaiming, and showed it to the students.

Finds of such obvious significance occur every day or so. Far more frequent are discoveries of less elegantly crafted artifacts, such as fragments of animal bone, seeds, and the other objects that must be collected, identified, and factored into an overall analysis of what ancient people were doing at Meadowcroft.

As of last summer Meadowcroft had yielded some 1,300 stone tools, 162 firepits, and 29 refuse or storage pits. There were also 211,818 bones or bone fragments from more than 65 animal species. Plant remains, including 24,640 hackberry seeds, have been attributed to more than 50 plant species.

Correlating mountains of such data from scores of strata covering thousands of years is the job of Dr. Joel Gunn, a University of Texas archeologist who specializes in computer analysis of data. Dr. Gunn, a codirector of the project with Dr. Adovasio, uses a computer terminal set up in a building near the dig to feed in new data and make analyses on the spot.

Dr. Gunn is now trying to correlate the changing diversities of plant communities represented at Meadowcroft with other evidence of global climate changes over the same period. Also under study is the spatial distribution of artifacts on each occupation floor.

"The way people use space in a given site is remarkably constant over time," Dr. Gunn said. For example, a pattern that persists at Meadowcroft from 9,000 years ago to 12,000 years ago is for the stone tools near the back of the shelter to be predominantly of locally available types of rock. Nearer the front, or exposed perimeter, of the shelter the stone tools are mostly made from rock types that could only have been imported from prehistoric quarries hundreds of miles away, including some in New York State and eastern Pennsylvania.

Dr. Gunn theorizes that this may reflect the classical division of labor along sexual lines. Men, being the hunters, tended to roam farther in pursuit of game than did the women who, being gatherers of plant food and restricted by children, confined their travels to the home region. Men, therefore, would have greater access to exotic rocks.

Dr. Gunn suggested that when both sexes were together at the rockshelter, the women may have concentrated their activities, using local rocks, at the back of the shelter while the men occupied the edge.

At 6 P.M. Dr. Adovasio dismisses the students on the dig and everyone hikes up the hillside to a tent camp set up near an abandoned building once used by miners stripping coal in the area. Both the dig and the camp are on the grounds of Meadowcroft Village, a restored nineteenth-century settlement whose owner, Albert Miller, discovered the archeological site in 1967.

While the students prepare dinner in the old mine house, Dr. Adovasio lifts weights for an hour. A terrarium in the mine house holds Dr. Adovasio's pet iguana, part of a camp menagerie that includes four cats, two chinchillas, and one dog.

"Archeology is the only thing I've ever wanted to do," Dr. Adovasio said after dinner. "I learned to read when I was four and was reading geology and history books when I was seven."

By the age of nineteen, Dr. Adovasio had worked on his first archeological dig and had begun pursuing degrees in anthropology at the University of Arizona, a major center of such research.

Although some archeologists believe that an understanding of the past sheds light on the present, Dr. Adovasio said, "There may well be some benefits for modern man in archeology, but I don't know of any."

"I do this because it's all I want to do," he added. "I want to take all the resources science can bring to bear and answer the questions we have about the people who lived here. Who were they? What were they doing here? How did they cope with their environment?"

At the end of the field season, Dr. Adovasio plans to shut down the Meadowcroft dig for at least a decade. Objects already recovered will keep laboratory analysts busy for many years, and undisturbed deposits, about 35 percent of the original, will remain for use when new methods of recovery and analysis are developed. Much of what archeologists do today at a dig was not done ten or twenty years ago.

"Also," Dr. Adovasio noted, "ten years from now we'll be asking entirely different questions about the past, questions we don't know enough to ask now. If we dug it all up now, we might destroy all the answers."

II
Earth, Sky, and the Subatomic World

ASTROPHYSICS, GEOLOGY, PHYSICS

Apparatus at Stanford where new types of atomic fragments have been detected. In it, positron-electron collisions occur, and energy and particles are produced.

Above: The Fermi National Accelerator Laboratory at Batavia, Illinois has a ring four miles in circumference, around which protons are accelerated to the highest energy achieved anywhere—more than 400 billion electron volts.

Right: Gordon P. Eaton, head of Hawaiian Volcano Observatory, stands on rim of Kilauea crater on the island of Hawaii.

If he can extend his gaze far enough using the powerful telescopes at Kitt Peak, Ariz., C. Roger Lynds hopes to be able to unlock clues about the fate of the universe.

6

EXPLORERS IN A SUBATOMIC WORLD

Walter Sullivan

"Oh my God!" said Wolfgang Kurt Herman Panofsky as he paced in front of the computer display inside Stanford's colliding beam ring. Normally Dr. Panofsky is highly articulate. On this occasion all he could say as he watched the fireworks displays on the computer screen was "Oh my God!" over and over. It was evident that the Stanford Linear Accelerator in California—the colossal experimental system over which he presides—was manufacturing short-lived particles of a kind never before seen.

It was one of those rare moments in high-energy physics when something entirely unexpected—something that "shouldn't be there"—makes itself known.

When it was learned that the same particle had been detected (in a very different manner) at the Brookhaven National Laboratory on the East Coast, cablegrams were sent to experimenters in Europe and the Soviet Union. Within days they had confirmed the finding. The Stanford group called it the psi particle, whereas at Brookhaven it was named the J.

Because of these discoveries and a variety of more recent findings, those working with the big accelerators—the big atom smashers—know they are on the track of something historic, even though it remains elusive. Theoretical concepts, such as the idea that "elementary" particles are really built of smaller "quarks" or that certain particles have a special property called "charm" have been tested, but some of the key predictions of those theories have not materialized.

Dr. Haim Harari, dean of the Graduate School of Science at the Weizmann Institute in Israel, in a recent group discussion at the Stanford center, where he also works as a theorist, likened the situation to that before magnetism was understood. If a magnet had positive and negative poles and the negative end was cut off, it seemed reasonable to expect that it would be entirely negative. Instead the fragment displays both positive and negative poles. Only when electric currents were understood did the explanation become clear.

Few physicists are lucky—or brilliant—enough to be in on discoveries as momentous as that of the psi or J particle. Is the rest all drudgery? Are the teams that do the research—theorists, computer specialists, experimenters, and engineers—so large that any feeling of personal achievement is submerged? And is the scale of the experiments so colossal and their complexity so great that their fruits seem artificial, with no link to our daily experience and needs?

There is no question that the experimental devices have become enormous. The Stanford Linear Accelerator (better known as SLAC) in Palo Alto, California, is a multimillion-dollar installation whose main facility for accelerating electrons is two miles long. Even larger is the Fermi National Accelerator Laboratory (known as Fermilab) in Batavia, Illinois. Around a ring four miles in circumference it accelerates protons (the nuclei of hydrogen atoms) to the highest energy achieved anywhere—more than 400 billion electron volts. It was built at a cost of $250 million and its annual operating budget is about $60 million.

It is also evident that the research teams are very large. The announcement of the psi particle discovery at SLAC was signed by thirty-five authors. The negative aspects of such bigness have been set forth in the Center Magazine by an avowedly disgruntled physicist, Dr. Robert Yaes of Memorial University in St. John's, Newfoundland.

"Bright young physicists," he wrote, "will be able to exercise little of their ingenuity, originality, and creativity when they are junior members of enormous research groups and are told exactly what to do." Many physicists, he added, "are now being subjected to a routine of boring, meaningless, alienating work."

A brief time spent with research teams both at SLAC and Fermilab has confirmed that a certain amount of drudgery is involved. But, particularly in the wake of the psi/J discoveries (a second particle was found at SLAC, plus the suggestion of a third), morale was high.

True, the research teams are large. Using a musical analogy, it can be said that a generation ago soloists were still in demand. Today, to break new ground, many physicists find that they have to play in an orchestra. Those interviewed at the two centers, however, seemed happy to be in the orchestra so long as they felt its performance was outstanding.

In this respect they are lucky. The experiments of significance today are larger in scale—and fewer in number—than before. While a brilliant physicist has no trouble finding a spot with one of them, those who are merely "competent" may be unable to do so. And, as noted by Dr. Burton Richter, head of the team that made the SLAC discovery, "Life is tough for those who start slow and hit it later."

The SLAC researchers proved to be living in two worlds. One is the familiar California scene of towering eucalyptus trees, flower-studded suburbia, and teeming freeways. The other is a world of bizarre concepts, huge and complex research tools, and long hours of experimentation. "We are married to that machine," one experimenter said, "and the wives at both ends get restless."

The concepts are bizarre in that they deal with phenomena remote from normal experience, such as the existence of antimatter, quarks, and charm, or the increase in mass that, because of relativity, occurs when matter approaches the speed of light.

In the SLAC experiments that revealed the new particles, tiny packets of matter (electrons) were collided head on with similar packets of antimatter (positrons). The positron is identical to the electron except that its electric charge is positive instead of negative.

When particles of matter and antimatter meet, both are completely converted to energy in a flash of gamma rays. But when they collide head on at high velocity, the energy of the collision also enters into the reaction. According to the formulations of Albert Einstein, energy can be transformed into matter and vice versa.

In the electron-positron collisions some of the resulting energy is immediately transformed into matter, producing a variety of particles. Because the colliding particles become more massive at very high energy, having two beams collide head on is far more effective than firing one beam at a stationary target.

The stationary electron has been likened by Dr. Sidney D. Drell of SLAC to a mouse that becomes an elephant when accelerated. Charging an elephant into a mouse would have little effect. Instead, at SLAC elephants charge one another —or, more precisely, elephants and antielephants. If one target were stationary, the accelerator, to achieve the same effect, would have to be 6,000 miles long.

The two-mile accelerator produces 360 bursts of high-energy electrons every second. Thirty of these are diverted into the colliding beam facility, some being used to produce positrons. The packets of electrons and positrons circulate in opposite directions inside the colliding beam ring, magnets keeping each compressed to toothpick size.

"The vacuum people are the real heroes," said Roy F. Schwitters of SLAC as he showed a visitor the ring. Even a few atoms of air in the tube through which the particles are flying can knock them out of orbit since, for hours, they whirl around the ring at almost the speed of light.

Because of the radiation hazard when the machine is operating, entrance to the building where collisions occur can only be gained after communicating by intercom with someone who has remote control over the gate. Each person entering must take a key from a series of locks on a panel. Only when all keys are returned and the area has been searched can the machine operate.

To record what happens when the beams collide, the collision area is enclosed in a massive series of detectors. When a newly created particle flies out, it first activates a triggering system that turns on the main detectors. These include spark chambers containing 100,000 electrically charged wires. The particle, which cannot itself be seen, leaves a trail of ionized gas along which sparks flash between oppositely charged wires. An eight-inch wall of iron around this complex is strongly magnetized. This bends the path of the particle in one direction if its charge is positive, in the opposite direction if its charge is negative, and not at all if uncharged.

Many of the collisions produce electron-positron pairs (called Bhabhas for a noted Indian physicist) and there are twenty-four "shower counters" to record them. Finally, above the iron box is the "muon tower." The muon is very similar to the electron, but some 200 times heavier, and it is short-lived. Muons have great penetrating capacity and thus escape the iron box to be recorded by spark chambers in the muon tower.

The output of all these detectors is digested by computers and displayed almost instantaneously on a scope. There, for example, a Bhabha (electron-positron pair) appears as two luminous lines radiating in opposite directions from the collision point. Or, occasionally, one sees the traces left by one or more heavy particles, such as the proton.

In the discovery experiment, it was when the energy of the collisions touched 3.105 billion electron volts that all kinds of streaks began flashing on the scope. Sometimes as many as fourteen prongs radiated from the collision sites. It was evident that a particle whose mass was equivalent to an energy of 3.105 billion electron volts was being mass produced and decaying into a shower of other particles.

Three computers are involved in these observations. They are programmed to reject spurious events, such as those caused by high-energy particles from space (cosmic rays). But a constant worry is that this digestion of data by a computer is introducing a bias into the results. As Dr. Richter put it, computers "are a depressingly important component of our life." Skill in programming and a feel for the ways in which such devices can mislead is therefore an important requirement for the new generation of physicists. Since, however, the data are all recorded on tape, events of special interest can later be studied free from computer-induced bias.

It is typical of such experiments that building the detection array took two years and required the skills of many specialists. Some members of the team joined it later and missed what, in some respects, was more challenging than the data collecting that followed. Such projects also run the risk that, after a prolonged construction period, the proposed experiment will have become obsolete.

A radical change has occurred since the early part of this century when the most fruitful physics experiments were done by one or two people on a tabletop at nominal cost. Instead of building accelerators, experimenters like Sir Ernest Rutherford used naturally accelerated particles. He discovered the atomic nucleus with a beam of alpha particles emitted radioactively by polonium. In Europe such experimentation remained the tradition until World War II. Spending large sums or employing engineers instead of one's own wits was looked down upon.

Nevertheless there were those who believed that to probe deeper into the atom would require far more powerful accelerators. When these—and the team approach to experimentation—began to appear, notably at Berkeley in California, some Europeans began to emigrate westward.

Today, as is evident at Fermilab, a further development has forced experimenters not only to work with big machines but, in a sense, to live with them. As a result, young physicists with recently acquired faculty appointments sometimes have almost no familiarity with their university and no teaching responsibility.

"I've never even seen Boston," exclaimed one member of the Harvard faculty. Typically the salary of such researchers is covered, all or in part, by a contract from the Energy Research and Development Administration (or later, the Department of Energy) and thus imposes a minimal burden on the university budget. In a few cases, too, members of the team commute to their university by air to teach.

While senior researchers stay closer to their home campus, one of the leading experimenters at Fermilab, Dr. Carlo Rubbia, professor of physics at Harvard, is sometimes referred to by students there as "the Alitalia professor." The reason is that he shuttles back and forth across the Atlantic, spending much time at CERN, the European atomic center near Geneva, as well as at Fermilab.

The experimenters are wedded to such centers because of dependence on the computers, data banks, and other facilities there. A decade ago the raw data consisted chiefly of particle-track photographs that could be brought home to the university for study. With the voluminous tape-recorded or photographic output of current experiments, this is no longer feasible.

The experiment being directed at Fermilab by Dr. Rubbia jointly with Dr. David B. Cline of the University of Wisconsin and Dr. Alfred K. Mann of the University of Pennsylvania is typical in this regard. As with the SLAC colliding beam, the purpose is to achieve clearcut results by using a beam of particles with minimal complexity. Electrons and positrons, for example, behave like points of electric charge. While they have mass, they apparently have no "size."

In the Harvard-Pennsylvania-Wisconsin experiment, a beam of neutrinos is used. They have neither mass nor electric charge and they interact with other particles only in terms of one force—the so-called weak force responsible for radioactivity.

Close to midnight on a recent evening, protons from the four-mile ring were being switched and shunted to feed sixteen separate experiments. The one dependent on a neutrino beam is designated 1-A, having been the first experiment planned for the laboratory as it neared completion several years ago.

One current goal of the experiment is to explore an astonishing report from India. Detectors 7,600 feet underground in the Kolar gold mines have recorded events taken to indicate that particles of great mass are penetrating the observation chamber from below and the sides, as well as from above. It is suspected that these are previously unknown particles generated inside rock surrounding the chamber by extremely high energy neutrinos, some of which have passed through the entire earth.

Since matter, on the scale of atoms, is largely composed of force fields, a neutrino, which ignores all but one of the forces in nature, can fly through matter with almost no likelihood of interacting. While the report from India is tentative, experiment 1-A is uniquely equipped to test it.

Every seven seconds the accelerator delivers to the 1-A experimental area a pulse of some 10 billion neutrinos. So great is this production rate that, even though neutrinos interact only rarely in the target material, there are enough interactions to simulate, in a matter of days, the effect postulated for the gold mine over years.

The starting point of the experimental array, whose development, like that at SLAC, took two years, is where protons extracted from the acceleration ring pass through twelve inches of aluminum. In doing so they collide with protons or neutrons in the aluminum, generating a shower of short-lived particles (kaons and pions). When these decay a fraction of a second later, the sought-after neutrinos are one of the products. To allow time for this decay the kaons and pions are allowed to fly down a vacuum pipe three feet in diameter and 1,300 feet long. Everything except the neutrinos is then stopped by walls of aluminum, steel, and a dirt bank, or "berm," 3,300 feet long.

The resulting beam, composed almost entirely of neutrinos and antineutrinos, passes through a succession of tanks filled with sixty tons of mineral oil. The tanks are divided into sixteen sections, in each of which devices record light flashes given off by the oil when electrically charged particles pass through it.

When a neutrino hits a nuclear particle in the oil, generating a shower of particles, the energy involved in the interaction can be determined from the resulting light flashes—an essential element for the analysis. One type of particle, the muon, is able to penetrate to the next stage of the array—a sequence of four massive, magnetized iron plates, each four feet thick.

Between them are thin plates electrically charged so that sparks leap between them along any path ionized by the passage of a muon. Through a system of mirrors these sparks are photographed so that the muon paths can be reconstructed.

On some fifty occasions in recent months, the production of muon pairs with opposite charge (counterparts of the electron-positron "Bhabhas") has suggested to the experimenters the existence of yet another heavy, short-lived product which they are calling the Y particle.

Setting such an experiment in motion is a team operation. In a typical warmup, Dr. T. Y. Ling, a Shanghai-born physicist, aged thirty-two, who obtained his doctorate at the University of Wisconsin, sits in front of a battery of television displays (there are eight in the control room). Through a keyboard and computer he can control the incoming beam of neutrinos.

Beside him is Dr. William T. Ford, aged thirty-four, educated at Carleton College, Princeton, and the California Institute of Technology. Through an oscilloscope he monitors the electronics. Behind him Dr. Richard Imlay, aged thirty-five, watches a printout of data. After studying at the universities of Maryland, Princeton, and Wisconsin, he has an appointment to Rutgers. Nearby Dr. Peter Wanderer, educated at Gonzaga University in his native Spokane, Washington, at Notre Dame, Yale, and Cornell, monitors other indicators. Now thirty-one years old, he is on the Wisconsin faculty, visits there "once in awhile," but has not taught.

On the night shift, with all running smoothly, only Dr. Peter Orr, a twenty-seven-year-old, tawny-headed Scotsman, is on duty, aided by Bradley Jensen, a University of Pennsylvania graduate student aged twenty-two. Over a loudspeaker they can hear the miniature thunderclaps of sparks within the detectors in the cavernous, darkened building next door, as well as the cameras clicking. The cameras are activated whenever there is an interaction—once every two or three pulses. After an eight-hour run some 1,500 photographs have been taken and by the next day they are ready for scanning.

This is done in a van parked between the abandoned house and barn of what had been one of the farms in this fertile area before Fermilab was built in the late 1960s. Some of the land is now being grazed by a herd of bison, and the center of the main, four-mile ring is being restored as native prairie.

The photographs show the tracks of particles generated by the collisions. Normally they are scanned by junior members of the team. However, when the chief experimenters, such as Dr. Cline, visit the site, they often spend an evening looking through the pictures for revealing phenomena that may have been overlooked.

Fermilab is an impressive installation. Its fourteen-story central building dominates the landscape much as does the Vertical Assembly Building at the Kennedy Space Center in Florida. Like the space program, its place in contemporary life has been likened to that of a medieval cathedral in manifesting the greatness and pride of an era. Certainly greatness is there. The world's leading theorists and experimenters are drawn to Fermilab by its unique facilities. Special measures are taken to encourage the exchange of ideas. Pads and pencils are provided on the large, round cafeteria tables. To allow for debates in transit there were even blackboards in the elevators—until it was found that they accumulated more graffiti than equations.

The medieval cathedrals, however, were built with enthusiastic support from the communities over which they towered. Fermilab encourages visits from the public, but those who depart with a real understanding of what is being done there are few. At SLAC a newly completed building was designed to be proof against sabotage—a holdover from student unrest a few years earlier.

After the discovery of atomic energy, costly physics research was often justified on the grounds that it might lead to other energy sources. Today, however, there is no real evidence that high-energy physicists are on the track of any such resource. Yet phenomena on the atomic level lie at the basis of all other natural processes, from brain function to the birth and death of a star.

At an after-hours get-together, some of the SLAC physicists conceded that on rare occasions they wished the fruits of their efforts were immediately applicable. But, as one of them put it, "what is relevant now will not necessarily be so in ten years." He and his colleagues are clearly looking far down the road.

The discovery of the psi (or J) particle won the 1976 Nobel Prize in Physics for leaders of the two teams involved in its detection: Burton Richter at Stanford and Samuel C. C. Ting of the Massachusetts Institute of Technology. Dr. Ting's group, working with a beam of protons carrying 30 billion electron volts of energy at the Brookhaven National Laboratory near Upton, New York, found the new particle by approaching from a direction almost opposite to that of the Stanford laboratory. Protons impinging on stationary protons produced large numbers of electron-positron pairs at 3.1 billion electron volts. As was immediately suggested by theorists, this was the first of a completely new family of particles, discovered in subsequent experiments in the United States and abroad.

7

PREDICTING EARTH'S MOST VIOLENT TENDENCIES

John Noble Wilford

HAWAII VOLCANOES NATIONAL PARK—The summit and slopes of the mountain are swelling slightly but menacingly. Deep underfoot, unseen but not undetected, molten rock churns with heat and gathering pressure. It seeps through subterranean crevices, probing with all its pent-up might for a way out.

These are the ominous signs, scientists say, that the world's largest active volcano, Mauna Loa, is building up for an eruption.

Based on the volcano's past behavior and the rumblings that register on the seismographs daily, sometimes every few minutes, scientists at the United States Geological Survey's Hawaiian Volcano Observatory are predicting that Mauna Loa should unloose a major flow of lava anytime between now and July 1978. They first made the prediction more than a year ago and have seen nothing in recent days to change their minds.

If they are right, a river of devastating lava would probably flow in the direction of Hilo, a city of 35,000 people about thirty miles from the summit. Several times in recorded history, lava has reached what are now the city's outskirts. Indeed, the entire island, the state's big island of Hawaii, is the creation of volcanoes that rise from the floor of the sea.

If the scientists are wrong and Mauna Loa contains itself

through July 1978, it would be no cause for relaxation, only confirmation once again of the fallibility of those who venture predictions about the Earth's more violent tendencies.

Gordon P. Eaton, the forty-eight-year-old head of the observatory, is the first to concede that predictive volcanology is far from a precise science.

"We don't have a theoretical model for predictions," he said. "We are more like social scientists. They see that people usually behave in a certain way and make predictions accordingly, but sometimes people act differently. That's the way it is in studying and predicting volcanoes."

Yet Dr. Eaton and the sixteen scientists and technicians who live in the shadow of Mauna Loa, elevation 13,700 feet, and work at the observatory on the rim of another volcano, Kilauea, elevation 4,000 feet, believe that by making a great outdoor laboratory of the two volcanoes, they are improving the odds in eruption predictions.

They have implanted forty-three seismometers in the area, the densest such network in the world. They have eighty-three tiltmeter stations in operation, sensitive to the slightest swelling or subsidence of the volcanic slopes. With laser instruments, they periodically measure precisely surveyed lines to see if there has been any expansion or contraction, particularly in the rift zones from which lava so often erupts. And they frequently remap the gravity and magnetic fields for signs of changes in the internal structure of the volcanoes.

Aside from providing clues to impending eruptions, such a comprehensive and systematic monitoring of the Hawaiian volcanoes is expected to add to the general knowledge of volcanoes. The Geological Survey hopes to adapt the monitoring methods developed here for use in the nation's other volcanic regions, in Alaska and in the Cascade Range of California, Oregon, and Washington.

This is what Thomas A. Jaggar had in mind when he

founded the observatory in 1912. The Massachusetts Institute of Technology geology professor had studied volcanoes in Italy, Japan, and Costa Rica, and thought there should be a permanent observatory.

The observatory, situated in the national park, is a modest complex of low metal buildings. From the windows Dr. Eaton can look out on the broad crater of Kilauea, an awesome sight that in 1866 moved Mark Twain to write: "Here was a yawning pit upon whose floor the armies of Russia could camp, and have room to spare."

Kilauea appeared serene. Some wisps of steam poured out of vents, but there were no pools of fresh red lava. But for the inside story, Dr. Eaton looked to the five drum seismographs standing near the window, rotating slowly, leaving squiggly traces, or to the bank of computers in the next room. The computers receive, digest, and store the radioed signals from all the observatory's seismic network.

"In the earlier days the observations were mostly visual," Dr. Eaton said. "But now we're so instrumented that we could function very well without being in view of the crater."

The night before, Dr. Eaton noted in checking the seismographs, a "shallow harmonic tremor" caused the ground to shake for about twenty minutes. If it had been a real shaker, alarms would have gone off in the homes of Dr. Eaton and several other scientists.

"We've had a number of tremors like that lately," Dr. Eaton explained. "It means that magma [molten rock] is moving somewhere down there. Mauna Loa had an eruption in July 1975 and many tremors before and after. Kilauea had a 7.2 [Richter scale] earthquake in November 1975. That so broke up the internal structure of the volcano that it is still in the process of reconsolidating and healing."

Every day the scientists and technicians trek out over their unstable terrain ("It's not your firm Manhattan schist," Dr.

Eaton said), installing or checking instruments and surveying for signs of crustal movement.

John Forbes goes early each morning to a vault built into the side of Kilauea's outer slope. He took a visitor through triple doors into an inner sanctum of seismology, where he changed the film for one of the key seismometers in the network.

Over on Kau Desert, a volcanic wasteland southwest of Kilauea crater, Mr. Forbes pointed to some of the newest rock on Earth. This is where the lava flows every few years, most recently in 1971 and 1974, killing off nearly everything and leaving a smooth, lustrous sheet of cooled lava, called pahoehoe. It is cracked here and there like city asphalt, and fern takes root in some of the cracks. It crunches underfoot like styrofoam and often collapses.

Sometimes, when the wind shifts, the air becomes acrid, as Mark Twain had observed. "The smell of sulfur is strong," he wrote, "but not unpleasant to a sinner."

On the floor of a smaller crater, Kilauea Iki, engineers from Sandia Laboratories in Albuquerque were testing the reliability of geophysical sensing devices in detecting buried molten rock. These could be useful in finding deposits of magma elsewhere in the world that could be new sources of energy. Scientists are hoping to convert the magma heat to electricity, as they can do with geothermal steam energy.

But no one wants to drill for magma until he is sure it is there. Since it is already known that at Kilauea Iki a 100-foot-deep pool of molten rock lies 150 feet under the crater floor, the crater has become a calibrating standard for testing the accuracy of detection instruments. If seismic or electric probes correctly spot the magma chamber there, they should be able to do so elsewhere.

The observatory's most immediate concern, however, has to do with eruption forecasting. And when an eruption alert is sounded, all members of the staff spring to action. They want to be ready to photograph the eruption, observe flow direction, sample lava and gases, and measure changes in the terrain.

There was a burst of seismic activity back in February, and one night it looked as if an eruption at Kilauea was imminent. Lennart Anderson, one of the scientists, remembers it well, for that was the night he engaged in what he called some "seat-of-the-pants seismology."

"We went into the area of the most pronounced shocks," Mr. Anderson said. "If you stood there, you didn't feel a thing. So I sat down on the road. There were four of us sitting there on the pavement. We agreed that all four of us had to feel something for it to be counted. It felt like someone was operating a piledriver off in the distance. We got so we could even get the direction of the shockwaves."

Later, Mr. Anderson went back and found that a new crack had opened in the earth not more than 100 feet from where he had been sitting on the road. "I don't know if I would do that sort of thing again," Mr. Anderson remarked.

In a more scientific vein, Mr. Anderson is experimenting with the use of electric currents and fields in exploring the volcanoes' changing internal structures. The work is based on the principle that magma is more conductive than solid rock and thus it should be possible to map its ebb and flow. Networks of permanently implanted electrodes are planned as yet another means of keeping watch on Kilauea and Mauna Loa.

Few areas of geology are as exciting as volcanology, according to Dr. Eaton. He grew up in Ohio, where, he said, "there hasn't been a volcano in hundreds of millions of years." He "got hooked" on geology at Wesleyan University, worked summers gathering data on farm wells in Connecticut, and earned his doctorate in geology and geophysics at the California Institute of Technology.

"So much of geology is working with a carcass, like a detective with a dead body," Dr. Eaton said. "You have to work backward from what you see and figure out how and why it happened. Here, you're observing the processes as they actually occur. It's a geologically dynamic situation, always active and always changing."

The dynamics of Mauna Loa are Dr. Eaton's most immediate concern. Over the last century the volcano has exhibited a typical eruptive cycle: a summit eruption, followed by another summit eruption, and then a rift outbreak on the northeast flank.

An eruption occurred on Mauna Loa's summit in July 1975, the first in twenty-five years. This ended the volcano's longest dormancy in recorded history; it usually erupts on an average of every three or four years.

Although the 1975 eruption was small by Mauna Loa's standards, a series of earthquakes followed the eruption, which suggested to observatory scientists that large volumes of magma were moving into the northeast rift zone, primed for a future eruption.

If the scientists should begin detecting a build-up in the number and intensity of tremors in that area, they believe they will have a few days to prepare for the predicted major eruption.

"There's no way we can stop a lava flow," Dr. Eaton told state and local officials at a planning meeting. "The only thing we can do is try to 'steer' it, or make it flow more broadly so that it will cool faster."

The observatory has four contingency plans. The first is to use a technique tried successfully by Dr. Jaggar in 1935—aerial bombing. The intent would be to disrupt the vents and channels to spread the flow, hoping it will cool and freeze in its tracks well before reaching Hilo.

Other methods include bulldozing walls of rubble to block the flow in narrow valleys and, if that should fail, pumping in ocean water to cool the lava. Iceland has used the latter method to good effect, Dr. Eaton said, but it would be more difficult in Hawaii because of the greater elevations from the sea.

As a last resort, Dr. Eaton said, Hilo would have to be evacuated.

Meanwhile, the volcano continues to swell and shake, and scientists of the Hawaiian Volcano Observatory maintain their vigil, keep an instrumented finger on the pulse of the volcano, and hope they have learned enough to be able to sound the alert well before Mauna Loa's next big show.

Kilauea erupted in September 1977, spouting a fountain of glowing lava. No one was injured. So far, Mauna Loa has remained quiet, though still menacing. Scientists of the United States Geological Survey began to hedge their prediction of a major eruption there by July 1978. But it is certain to happen, they said, sooner or later.

8

AN ASTRONOMER LOOKS BACK TO SEE AHEAD

Walter Sullivan

KITT PEAK, Arizona—By looking back as far as possible toward the beginning C. Roger Lynds hopes, with one of the world's most powerful telescopes and a novel observing system, to determine the fate of the universe.

If he can extend his gaze far enough into space—and therefore back in time toward the infancy of the universe—he may be able to tell whether its present expansion will continue forever or has been slowing enough, over the past few billion years, to indicate ultimate reversal and collapse.

Other astronomers have tried to resolve the question in a variety of ways. Some see evidence for eternal expansion. Others are unconvinced.

Dr. Lynds is attacking the problem from a new angle, using an observing method dependent on an image intensifier, a vidicon tube that converts the image to electric impulses, two computers to process them, and television displays.

Specifically, he is trying to determine whether extremely distant galaxies of a certain type, viewed as they were billions of years ago, are brighter, dimmer, or differently shaped from those of today. It has been the possibility of such "evolutionary effects" that has clouded many past efforts to answer the key question: to what extent is expansion of the universe slowing?

"I am by no means convinced we'll find the answer," he says with a quizzical grin. But, he adds, the stakes are so high that the effort is more than worthwhile. Included in those stakes is a deeper understanding of how the universe and its galaxies have evolved.

Dr. Lynds' tool is the main telescope of the Kitt Peak National Observatory with a mirror four meters (158 inches) in diameter. Only the five-meter (200–inch) telescope on Mount Palomar in California and the new Soviet six-meter (236–inch) reflector in the Caucasus are larger. A twin of the instrument here has been constructed at Kitt Peak's sister observatory atop the Chilean Andes at Cerro Tololo.

About 5 P.M. Dr. Lynds, aged sixty, jumps out of bed in the dormitory marked "Day sleepers—quiet please" and eats a hearty meal in the observatory dining hall. It is then a short drive to the telescope which is supported by the highest mount in the world—comparable to a nineteen-story building. Its height is designed to minimize distortions in the telescope's field of view caused by turbulent air sweeping over the mountain.

The mount is only a few yards from the rocky summit, 6,875 feet above sea level. As Dr. Lynds and his coworkers park their car on a terrace carved from the rock, the setting sun glints on more than a dozen smaller domes rising from vantage points on the mountain like gleaming white mushrooms. While most belong to the Kitt Peak National Observatory, others are operated by the University of Arizona and the National Radio Astronomy Observatory.

Fifty-six miles to the northeast, a panoply of lights marks the location of Tucson. Otherwise, as the night darkens, few lights threaten to interfere with sensitive observations for the mountain is within a sparsely inhabited reservation of the Papago Indians, from whom the site is leased.

Twelve miles to the south, rising above the mountainous

skyline like a giant thumb, is Baboquivari, a 7,730–foot mountain which, according to Papago legend, is the center of the universe, inhabited by its creator. Not far beyond lies Mexico.

With a passkey Dr. Lynds opens the door and, with his companions, rides an elevator to an intermediate level, where there are sleeping accommodations for those who do not wish to leave the site, as well as two observing computers. A second elevator then carries the group to the control and observation room. Dr. Lynds tosses a heavily padded, quilted jacket onto a chair, for the room is heated.

The observing system is such that he never peers directly through the telescope. On a typical night he may not even see the giant instrument, mounted in a frigid dome a few feet away and aimed through a slot at the open sky. He sits watching television displays and computer printouts. These indicate where the scope is aimed and, when the aim is changed, rumbling manifests rotation of the dome to keep its open slot above the slewing telescope. The corridors moan and whistle as wind blows through cracks in the structure, which also trembles gently, but the giant telescope is mounted to remain isolated from such vibration. It is so perfectly balanced that even when its 375 tons are swinging to a new target one feels no quiver.

The first step tonight is to calibrate the sensitivity of the system by aiming it at an object of known brightness—the edge of a globular cluster of stars. The cluster lies on the outskirts of the galaxy to which the sun and its planets, as well as the stars of the Milky Way, belong. From strange patterns on a television monitor it soon becomes evident that something is amiss. A call for help to Tucson brings one of the fifteen specialists who developed the system in its improved form—in use, so far, for only a few nights. There is time for repairs since the moon, whose light impedes observing, will not set until midnight.

A Scarlatti sonata, in tape-recorded performance by the harpsichordist Ralph Kirkpatrick, tinkles lightly, helping maintain a serene atmosphere in the observing room despite the difficulties. Drs. Lynds and Vahé Petrosian, theorist from Stanford University, go over plans for the night. Paul A. Scott, programmer of the computers, checks them out.

Shortly before midnight there is a partial exodus for the "noon" meal. The car creeps down the road, keeping clear of the precipice that drops away to one side. Only parking lights are permitted lest a headlight flash into a telescope mount. The moon is almost down and it is very dark.

Dr. Lynds tells of his origins and his early interest in music. His father was principal of a rural Kansas high school and, to flesh out the school orchestra, young Roger had to play the horn, cello, and violin. Now he and his wife, Beverly, herself a distinguished astronomer and, at times, acting director of the observatory, have a harpsichord in their home. They find the difficulty of keeping it in tune a handicap.

In explaining their project Dr. Petrosian notes that galaxies occur in many forms, such as the spiral structure of the Milky Way Galaxy and the elliptical shape of the brightest ones. He likens the many varieties to melons, bananas, cucumbers, and oranges. It is by observing numerous giant ellipticals—the watermelons among galaxies—that Lynds hopes to nail down the extent to which expansion of the universe has slowed.

The discovery that the universe is expanding was made by Edwin P. Hubble in the 1920s when he observed that virtually all galaxies beyond our own, the Milky Way system, are moving away. The greater their distance, he found, the more rapid their rate of recession. The galaxies are moving apart like particles in an expanding cloud of gas.

His estimates of relative distances were based on the dimming of light from the galaxies. He determined their rate of motion away from the earth by the extent to which characteristic wave lengths of their light were lengthened by such motion—that is, shifted toward the red end of the spectrum. This "red shift" is similar to the lowered pitch of a horn on a receding vehicle.

To learn the expansion rate more precisely it was necessary to use only galaxies of similar intrinsic brightness—constituting what astronomers call a "standard candle." If their intrinsic luminosity was the same—like so many 100–watt light bulbs—then the extent to which their light was dimmed would be a more uniform index of their relative distances.

To this end Dr. Allan Sandage, using the 200–inch Palomar telescope to observe the giant ellipticals, has tried to trace the expansion far enough into the past to see if it has slowed substantially. One problem has been seeing far enough for the effect to become evident. Another has been the difficulty of obtaining accurate measurements of relative brightness. And finally, it is suspected that the intrinsic brightness may have changed. Perhaps when such galaxies were young and full of newly formed stars they were brighter —1,000–watt bulbs instead of 100–watt bulbs—and unsuitable as standard candles.

It is to learn if such changes have occurred that Dr. Lynds, in collaboration with Dr. Sandage, is surveying the brightness of giant ellipticals near and far. If it has remained the same since galaxies first formed, the brightness of a point in the image of a nearby galaxy should be the same as that of a point in a very distant one, although a correction must be made for an effect of relativity on light from a source moving away at extremely high velocity. Apart from that, a distant 100–watt bulb looks dimmer only because it occupies a smaller part of the field of view. The area is less according to the square of the distance and so, therefore, is the total brightness.

Traditionally observations of the most distant galaxies and what seem the even more distant quasars are made by long photographic exposures in which a high-precision tracking system keeps the telescope aimed squarely at the object. Photographic emulsions, however, are not a reliable indicator of relative brightness. The system in use here records numerically the brightness of each spot or "pixel" of the image, making the latter amenable to computer manipulation much like the pictures transmitted by the Viking landers on Mars.

The system is so sensitive it could be matched only by photographic exposures eight to ten times longer. One computer digests the data for subsequent analysis—again with computer aid—at the Tucson laboratory of the observatory. The other computer processes the impulses for direct television display.

Tonight, with the observational devices again functioning, William Halbedel, the telescope operator, types instructions onto a keyboard to swing the scope and dome into position for the first observation. Stellar images streak like shooting stars across the televized acquisition screen in front of him as the field of view moves across the sky.

To remove the effects of twinkle, caused by atmospheric turbulence, a star is repeatedly observed at the same time as the galaxy. The star is assumed to be a point source of light, jumping around the field of view because of atmospheric effects. Through a process called deconvolution, repeated observations of the star make it possible to narrow its image to a point, and the same corrections can then be applied to the galaxy, whose light has been passing through similar turbulence.

Normally a star in the same field of view is used, but the vidicon field is so narrow that sometimes there is no suitable star within it. Tonight, therefore, the telescope is automatically swung back and forth very slightly between a galaxy and a star still close enough for twinkle correction.

Each recording period or "read" lasts 1.6 seconds and recordings are made successively through as many as four filters, passing red, green, blue, and white light. For each filter there are 32 "reads" on the star and 128 of them on the galaxy, each series being superimposed.

So sensitive a system picks up many stray light signals and the whole monitoring scope "swims" with such extraneous "noise." It is like looking down from an aircraft at chaotic ocean wave motions. The noise, however, largely vanishes as, with the help of the computer, successive images are superimposed. With each superimposition, images of the stars and galaxies become more intense whereas spurious spots, lacking reinforcement, fade away.

The stars being observed alongside galaxies are a few hundred light-years away—next-door neighbors on the scale of the universe. Some of the galaxies, almost starlike in appearance, are billions of light-years away; that is, their light has taken that long to reach the telescope, traveling at 186,000 miles a second, and they are being seen when the universe was relatively young.

Once the galactic images have been fully processed, it is hoped that analysis of their brightness, from midpoint of the image to its faintest outer edge, will reveal any systematic differences between the very distant ones (observed in their youth) and those close at hand.

The first hint of dawn washes out the observations, but to make possible later correction for distortions across the field of the recording system the scope must be aimed at a flat, uniform field of view, namely the inside of the dome. It is

7:30 A.M. when Dr. Lynds "calls it a day" and, in the mess hall, shares breakfast with those living normal hours. He notes that some astronomers, such as Dr. Sandage, insist during the daylight hours on developing films exposed the night before. While, in this case, no film developing is involved, these winter nights are long, Dr. Lynds says, "and you end up getting very little sleep."

III
Invention

ELECTRONICS

Alec Feiner at his desk at Bell Telephone Laboratories.

9

THE FERREED CROSSPOINT: A REVOLUTION IN TELEPHONES

Victor K. McElheny

NEW YORK—Alec Feiner is a member of the much-honored but poorly understood profession followed by Eli Whitney of the cotton gin, Samuel F. B. Morse of the telegraph, Alexander Graham Bell of the telephone, and Thomas Alva Edison of the electric light.

Mr. Feiner is an inventor. In fact, he produced one of the most successful inventions in the history of that hothouse of invention, Bell Telephone Laboratories, founded in 1925.

At the New Jersey-based Bell Labs, birthplace of radio astronomy and the transistor, Mr. Feiner invented the basic electrical contact, known as the "ferreed" crosspoint, used in the faster, higher-capacity electronic telephone exchanges being installed every few days across the country, alongside older equipment in the Bell System.

The ferreed that Mr. Feiner invented is a switching mechanism involving two tiny metal reeds in a glass capsule that also contains two coils of wire. The coils can be magnetized in the same direction to bring the two reeds together, or in opposite directions to drive the reeds apart.

Much faster than previous generations of dial telephone switches, and yet, like its predecessors, involving a firm metal contact for opening up a circuit, the ferreed is the means by which about one-fifth of the Bell System's 70 million telephone lines were linked to the others by 1977, a dozen years after the first commercial Bell System electronic exchange was put into service in Succasunna, New Jersey.

Without the ferreed, the United States telephone system probably would have been forced to adopt more complex methods of harnessing electronic technology to handle the growth from 40 million telephones in 1950 to more than 160 million in 1977.

In 1969, a decade after his invention of the ferreed, Mr. Feiner became the head of a whole team of inventors at Bell Labs in one of the most competitive of all communications businesses, the provision of private branch telephone exchanges.

Unlike his famous predecessors, instead of inventing whole systems or whole industries, Mr. Feiner has invented components of systems.

Instead of working alone, in a basement or attic, Mr. Feiner works in a huge, glass-enclosed building, not too different from structures at some of the world's largest universities. He works for an industrial research organization that regularly spends about 2 percent of the revenues of the American Telephone and Telegraph Company, the Bell System, that reached nearly $33 billion in 1976, and employs more than 16,000 people, including 2,000 doctoral degree holders.

The Vienna-born Mr. Feiner, who grew up in Krakow in Poland and spent four years of his youth in Soviet prison camps during World War II, is not one of the doctorate holders. For the last quarter-century, or since the age of twenty-five, he has been so engrossed in a succession of inventions, contributing to a kind of electronic revolution gradually pervading the telephone system, that he never had time to write a doctoral thesis. Just a few years ago, he married a South African-born physician, and is the father of a little son and daughter.

When the slight, curly-haired Mr. Feiner was interviewed in his office at Holmdel, New Jersey, Mr. Feiner said, "Treat me as an average inventor."

Many would regard him as an above-average inventor, since several of the more than thirty inventions he has made have been applied commercially. The usual average for practical adoption of inventions is less than one in a hundred.

He discussed his work in a carpeted office furnished with a modern couch, desk chairs, and bookshelves, a chart showing production totals of the different systems to which his laboratory has contributed, and two small prints of scenes in the French city of Rouen.

Despite differences from the inventors of less complicated days, Mr. Feiner resembles them in at least one respect, his passion for elegantly simple solutions to technical problems.

Although he was "never afraid to go as deep as I had to into basic physics" to attack a problem, Mr. Feiner said that his "motivation always was practical."

He continued, "I wanted to do something useful and new and simple. To me, simplicity is the most elegant thing. To reduce the most complex to the most simple has great importance for me."

Simplicity would seem a virtue to the telephone engineers of the 1950s, as they sought to develop fast and flexible electronic switching for a rapidly increasing volume of local and long-distance telephone calls.

Although the older systems were less flexible and slower than electronic systems could be, they were reliable, simple, and durable.

New systems would not so much displace them as struggle for an economic place alongside them.

The engineers knew that the existing ideas would have to

be "completely rethought," Mr. Feiner said. "It was a revolution, no longer an evolution."

The basic concepts of a much faster and more flexible, computer-controlled system of switching telephone calls were laid down in 1951 by Chester E. Brooks, a Bell Labs systems engineer. The work of bringing the electronic concepts to their first field trial, in Morris, Illinois, in 1960, was led by Clarence A. Lovell, a pioneer in work on analog computers, and Raymond W. Ketchledge, who became an executive director at Bell Labs' center for electronic switching development at Indian Hill, Illinois, west of Chicago.

To capture the speed and flexibility inherent in electronic devices for the telephone system, the controls had to be centralized in a "superfast brain that oversees all," as Mr. Feiner described it. And because that brain could "crash" or break down like any other computer, it was given a twin to "continually compare notes" and take over in emergencies.

The "real time" computers to be harnessed to the telephone system, Mr. Feiner noted, were not the same as the devices cranking out mathematical solutions for scientists at universities. The telephone computers were supposed to be setting up telephone connections, hundreds of them every minute.

Mr. Feiner approached his own central contribution to this technology indirectly. In this, he resembled such inventors as James Watt, who got started improving steam engines when there was trouble with the old-style model used in a university science course, or Alexander Graham Bell, who stumbled across the reality of electrical speech transmission in the course of experiments on the telegraph.

Mr. Feiner did not set out to develop a general cross-point for electronic telephone exchanges. At first, his problem was how to help save the scarce metal copper in an electronic network.

It seemed that there might be economies in handling a group of subscribers' lines through a satellite exchange, or "remote concentrator," rather than running a separate copper connector from each phone to the central exchange itself.

If such a remote concentrator were used, how could connections be set up, with a minimum of complexity and a minimum of demand for electricity? The electricity would be supplied from telephone sources—and most of it would have to be reserved for the big electronic brain at the telephone exchange.

In previous generations of switching equipment, connections were made by metal bars that jumped or flipped, opening or closing circuits under the control of magnets. The switches themselves were sturdy and the related circuitry could handle the relatively robust current needed for ringing a telephone. The trial electronic exchange in Morris, Illinois, relied extensively on gas-filled tubes typical of the early computers, and needed special new telephones because the electronic equipment could not stand a normal ringing current.

One morning in the late 1950s, after months of tests and "thought experiments" with his colleagues, Clarence Lovell, T. N. Lowry, and P. G. Ridinger, Mr. Feiner arrived at his office at Whippany, New Jersey, with a possible solution to the problem of a cross-point that would be sturdy and simple, and yet low in power demand and fast-acting enough to be compatible with an electronic system.

It was cold that morning, Mr. Feiner recalled, and so he wore his overcoat to work. He was so intent on explaining the concept of the ferreed to his colleagues that he forgot to take the overcoat off until lunchtime.

The night before, after months of intense concentration, he had thought through basic ideas leading to the tiny glass capsule containing two slim, magnetic reeds and two parallel coils of wire. By creating parallel magnetic fields around coils with a tiny circuit, the two metal reeds could be slapped together. By reversing one of the magnetic fields, the reeds could be driven apart.

The device worked so successfully that it was adopted as the basic switching mechanism for the electronic telephone exchanges themselves, rather than in remote concentrators.

The success brought enormous relief to telephone system engineers. They had found that the concept of a computer-controlled exchange, using "stored programs," had worked well in Morris, Illinois, but there was one problem. The equipment demanded installation of special telephones. Would the electronic revolution demand millions of new telephones? The invention of the fast but sturdy ferreed removed this economic obstacle.

Since 1965, hundreds of so-called Number 1 E.S.S. exchanges based on this technology have been installed in the nation's central business districts, including dozens in New York City, and several hundred more Number 2 and Number 3 E.S.S. exchanges have gone into suburban and rural areas.

One key result of the new electronic switching has been much greater flexibility in the type of service that can be offered to an estimated 12 million private branch or Centrex telephones in the United States, which has roughly one-third of the world's telephones.

The business of supplying such equipment, on which Mr. Feiner's laboratory concentrates, was opened to Bell System competitors by a 1969 decision of the Federal Communications Commission.

Like many others who are impelled to understand things in a new light, Mr. Feiner experienced a great deal of turmoil and risk in his early life. The German invasion of 1939 drove his family from Krakow eastward to the shelter of Soviet prison camps, including one near Archangelsk in the Arctic.

The chaos of the German retreat from the Ukraine in 1944 allowed a chance of escape westward, dodging soldiers, partisans, and bands of homeless, murdering children. Back in Krakow, the family fled again, to Vienna, because of food shortages, pogroms, and tightening Communist control over postwar Poland.

After four years' studying electronics at the Technical University in Vienna, Mr. Feiner got a United States visa, entered Columbia University, added English to his languages, received a master's degree, and took a course taught by a Bell Labs staff member who encouraged him to seek employment there.

It was an "extremely lively" time to join Bell Labs, Mr. Feiner recalled. "The electronic era in switching was dawning."

Now, he said, a young designer of a new electronic system "can pick out many of the devices he needs." In earlier days, such electronic memory devices as magnetic cores or twisted wires "hardly existed," he said, and "It was not quite clear what to do; there was no idea that quite fitted."

His ferreed idea fitted quite well, and like many other pioneers in electronic switching development, Mr. Feiner was given increasing responsibilities in developing the systems that flowed from the first innovations.

Mr. Feiner became the supervisor of a largely young staff working in movable-wall laboratories along corridors that end in balconies overlooking huge open spaces within the huge glass building at Holmdel. "I love to interact with my young people," Mr. Feiner said. "In a sense, I try to invent through them. I try to help them obtain information they need, suggest a line of research, suggest a person to talk to."

The younger engineers and scientists, he said, "seem to get brighter every year."

"They all know how to use computers very creatively," he

added. "They are getting a more appropriate education. Twenty years ago, it seemed sometimes as if the professors didn't know what to teach us."

In the years since 1953 at Bell Labs, Mr. Feiner has come up with several inventions that have been applied commercially. One of these, which he invented overnight "on a dare," is a conference-call circuit that bears his name.

To Mr. Feiner, the process of inventing is difficult to describe. He says that it involves "extremely intensive thought, a certain pressure on your brain." Much of the work is subconscious, Mr. Feiner thinks.

"It is as if there are two of us, one inside the other," he said. "All of a sudden, a possible solution bursts into conscious thought."

"Invention springs from purpose," Mr. Feiner observed. "A lot of hard work precedes it. You have to give your brain what it needs. There is intensive thought and intensive interaction with others." An inventor, he said, "must be prepared to come up with completely foolish ideas."

"The hardest thing is to have courage in your own mental powers," he believes. "So the hardest invention is the first one. Later on, you have the fear that you will lose that creative power."

IV
Exploration of Life—I

BIOLOGY, CHEMISTRY, GENETICS, PHYSIOLOGY

Male moths, such as giant silk moth, have antennae tuned to chemicals of females.

Dr. David Karnosky at work at the Carey Arboretum in Millbrook, N.Y., where he is trying to breed a perfect elm tree with the beauty and shade of the American elm, but resistant to the Dutch elm disease.

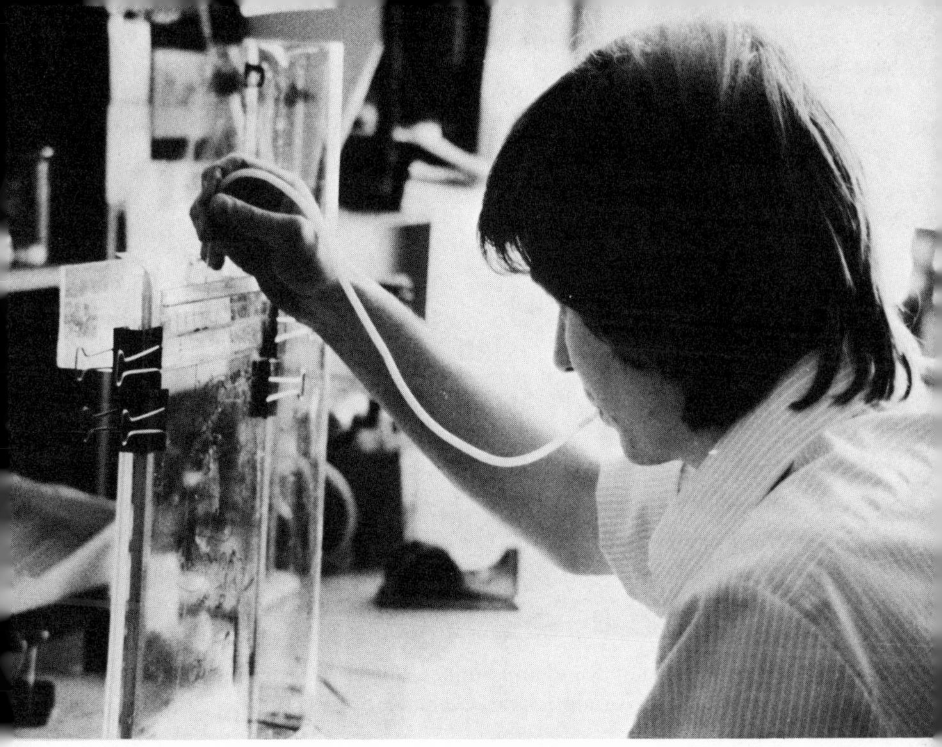

Left: Richard W. Robinson with experimental tomato plants at Geneva, N.Y.

Above: Dr. John C. Fiddes of the Laboratory of Molecular Biology in Cambridge, England, uses suction to lift a solution of DNA segments and deposit them in a device for electrical analysis (electrophoresis).

Right: Prof. Thomas R. Odhiambo, standing, with researcher at insect physiology center at Nairobi.

Dr. A. Stanley Rand and Stella Guerrero examining a small lizard on Barro Colorado Island, Canal Zone.

Dr. Ralph A Nelson, right, and members of his scientific team taking blood from one of the bears to use for comparison with later samples to be taken during the hibernation period.

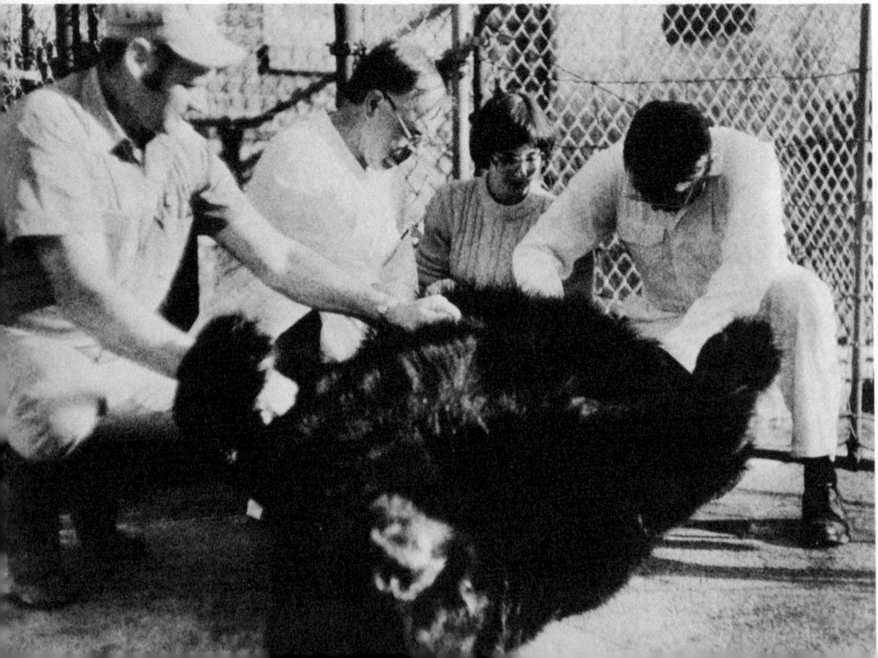

William Nunn Lipscomb, Jr., with a molecular model in his laboratory.

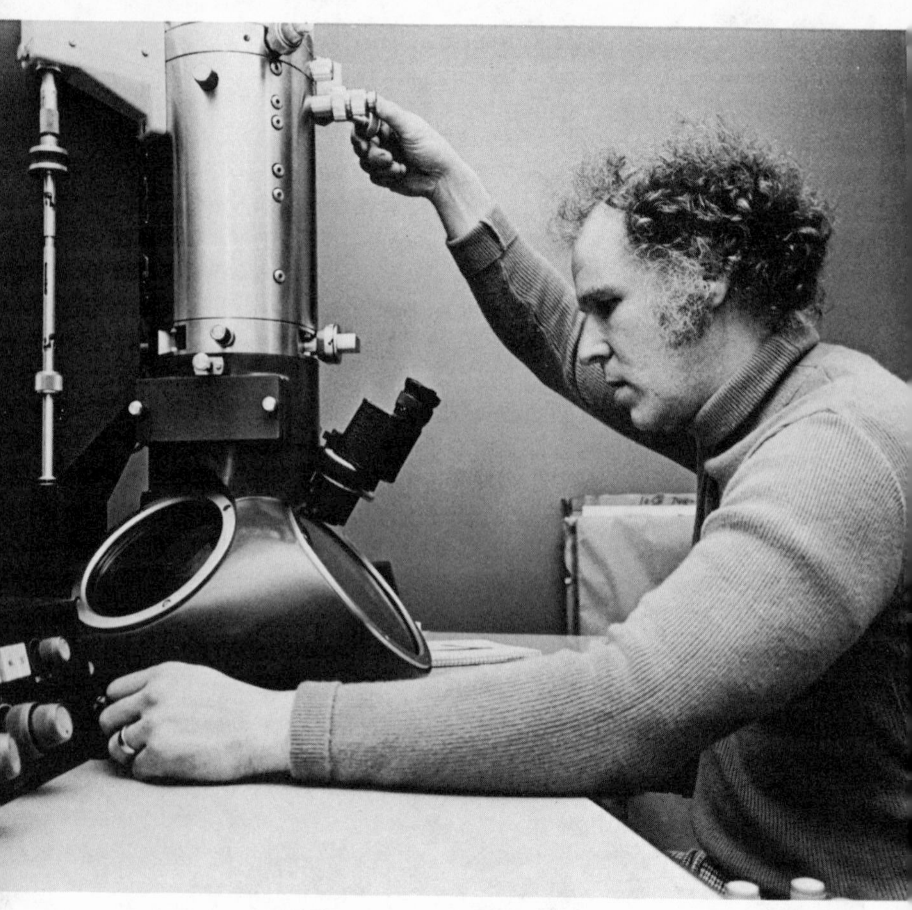

David M. Phillips, a researcher at the Population Council, at work in his lab with an electron microscope which he uses to study forms of sperm.

10

THE CHEMISTRY OF SEX AND PEST CONTROL

Jane E. Brody

GENEVA, New York—Redbanded leafroller moths are flying this month through the apple orchards of upstate New York, their antennae tuned for their main task in life. The male moth, which lives for only about two weeks and does not feed, must "sniff out" a mate to produce the eggs for the next generation of caterpillars that threaten the state's $50-million-a-year apple crop.

The caterpillars roll themselves into apple leaves and, well protected from pesticide sprays, champ on the leaves and eat the skin off nearby fruit, causing scabs and rotting.

This is also the month that Dr. Wendell Roelofs and his colleagues at the New York State Agricultural Experiment Station here are racing through the budding orchards hanging traps baited with the various chemicals in the sex scent, or pheromone, that the female moths use to attract the males.

The traps are part of a test of a new approach to pest control that may ultimately solve the apple growers' problem with leafrollers and other orchard pests without the extensive use of costly, highly poisonous pesticides that disrupt the balance of nature and sometimes create more problems than they solve.

Instead of using pesticides, Dr. Roelofs and others hope to control some insect pests by spraying manmade versions of the insect's sex attractant chemical so that the males will be unable to locate the real females for mating.

The trapping tests now in progress are the culmination of work that began in 1968 when Dr. Roelofs first isolated and identified the main attractant chemical used by the female redbanded leafroller. He subsequently unravelled the intricate chemistry of the sex pheromone by a controversial technique of measuring the electrical responses of detached male moth antennae to various substances extracted from the female's scent gland.

In the last four years, the ingenious but "baby-simple" technique, called electroantennography, has earned the thirty-six-year-old scientist an international reputation among entomologists, although he himself has never taken a course in entomology or biology.

Since 1971, Dr. Roelofs has used the EAG technique to determine quickly and accurately the chemical nature of the sex attractants in about half of the fifty species of moths whose signalling chemistry is now known.

In 1973, the Entomological Society of America awarded him its top honor for "outstanding work in economic entomology," although many insect experts remain distrustful of his method.

Despite the skepticism, hardly a week goes by at Dr. Roelofs' laboratory without the arrival of a small perishable package from some other state or country containing the pupae of a troublesome moth for his team to analyze.

With all his accomplishments, Dr. Roelofs confines his work to a regular eight-hour day including an hour for lunch, which in winter he spends playing basketball at the "Y." He believes the best ideas come to him, not when he works fourteen or fifteen hours a day, but "when the pressure is off —while I'm cooking or playing tennis."

Dr. Roelofs, a chemist by training, got into the biology-dominated field of entomology a decade ago because he was stimulated by the challenge of a new science.

"Pheromones were just beginning in 1965," the blond, muscular scientist recalled in an interview, "and I was excited by the idea of coming into a new field and setting up my own laboratory," now housed in a modern, six-story entomology building then under construction at the experiment station.

"When I arrived here, there was no chemistry instrumentation at all in the entomology department." Now he has four gas chromatographs and a mass spectrophotometer among the tools used by the three entomologists, two chemists, and two technicians in his laboratory.

While the new building was being completed, Dr. Roelofs combed the scientific literature, reading eagerly about insects and the various methods that had been used to study their communications systems.

The old ways of biologists—bioassay techniques that are still widely used—seemed too tedious, expensive, and imprecise. They involve exposing thousands of live males to extracts from thousands of females, and repeated testing of the moths' behavioral responses as the extracts are further and further refined chemically.

These tests have to be conducted at the right temperature and time of day or night when the males are naturally responsive to the females' sex signal. Sometimes a combination of chemicals is needed to get the behavioral response, making it impossible to assay separately the different active compounds in the sex pheromone.

To identify the redbanded leafroller moth's main pheromone chemical with this approach, Dr. Roelofs had to use extracts from the scent glands of 40,000 females. The work took him two years.

He figured there had to be an easier way—"we try to do

everything around here as simply as possible"—and his search of the literature uncovered one.

In the 1950s, a physiologist named Dietrich Schneider at the Max Planck Institute in Germany began studying moth antennae to learn how they perceive odors. He did this by hooking up the antennae of male silkworm moths to electrodes and recording the signals given off as the antennae were exposed to odors from the females.

Then in 1969, British scientists reported using the electroantennography method in an attempt to isolate the pheromone of a cotton pest. They passed scent gland extract through a gas chromatograph to separate component chemicals and tested the response of the antennae to the various substances that came through.

Dr. Roelofs decided to try it, and a few hundred insects and four days later, he had identified the attractant of the codling moth, a major apple pest whose larvae become the proverbial worm in the apple. Using traditional methods, others had gone through half a million insects in five years and still had not identified the pheromone, Dr. Roelofs said.

Obtaining pheromone for analysis by the new method starts with mass-rearing the insect in question, the responsibility of Frances Wadhams, a technician with "a green thumb for rearing insects," who keeps reproductive colonies of fifteen to twenty different insect species going at any one time.

"Our whole group hinges on Fran's ability to keep all these insects at different life stages," Dr. Roelofs said, with no small measure of appreciation for her talents at producing the thousands of adults of various species needed each week for pheromone studies, while maintaining others in "holding cultures" until they are again needed in large quantities. Each insect has its own peculiar dietary and environmental requirements, and it is Mrs. Wadhams' job to figure out what each needs in order to thrive, reproduce, and remain free of disease.

Next comes the painstaking separation under a microscope of the male and female pupae (the stage between larva and adult) by counting the number of segments on the body of each pupa. Dr. Roelofs explained that if the two sexes are allowed to emerge as adults together, they often mate right away and the amount of pheromone in the female is greatly diminished.

When the adult females emerge, Tom Baker—"our expert at squeezing females"—and Kathy Poole go to work. Under a microscope, they pinch the abdomen of the insect (the whole leafroller is only half an inch long), squeeze out the pinhead-sized pheromone gland, and snip it off into a solution that extracts the fatlike scent chemicals.

This extract is then run through the gas chromatograph, which separates chemicals on the basis of their molecular weight, and once a minute the fractions are collected in tiny tubes. The thirty or more tubes are then subjected to electroantennography analysis.

As Dr. Roelofs has refined it, the technique consists of snipping off an antenna from a live male moth and placing it in a dish of salt water, with the tip of the antenna sticking out of the water. One silver electrode is also placed in the dish. The second electrode is inside a tiny tube filled with salt water, with the end of the tube just touching the tip of the antenna.

By using salt water as the conductor of electricity, Dr. Roelofs avoids the arduous task of attaching the wire electrodes directly to the tiny antenna.

Then tiny puffs of air containing the different chemical fractions from the gas chromatograph are blown across the antenna and the electrical response of the odor-sensitive organ is recorded on an oscilloscope. When a particular fraction elicits a strong response, Dr. Roelofs selects it for further fractionation and analysis of its likely chemical structure.

"The beauty of the antenna is that its responsiveness is not restricted to a particular time or temperature, as is the brain-directed behavior of the intact insect," Dr. Roelofs remarked.

The molecular analysis is greatly aided by sophisticated chemical guesswork based largely on how the antenna responds to Dr. Roelofs' extensive "library of compounds" of known structure. The antenna responds to chemicals that are closely related to the structure of the insect's natural attractant chemicals, and this helps Dr. Roelofs to narrow down the possibilites.

The general structure of the pheromone chemicals is also deduced on the basis of when the active chemicals come off the chromatograph compared to the standard compounds in Dr. Roelofs' library.

The active fractions from the insect pheromone are subjected to further separation techniques and repeated EAG analysis until Dr. Roelofs is confident that he has pure chemicals of known structure. He then does a rigorous chemical proof of the structure.

The final chemicals are run through a brief bioassay to be sure the scientists are indeed left with substances that stimulate sexual behavior in the male insects.

Then comes the task of synthesizing the pheromone chemicals, a job that Dr. Roelofs frequently farms out to another laboratory that can produce manmade versions in a few weeks.

The final, crucial, and most time-consuming step is the field tests, which usually take two years, to determine the attractability of the synthetic pheromone to real insects in real situations.

Dr. Roelofs has so refined the antenna technique that he can sometimes identify the chemistry of the pheromone using as few as ten or twenty insects, containing all together less than a nanogram of attractant. In one case, he was able to identify the attractant chemical within half an hour of the insect's arrival at his laboratory.

Analysis by the antenna technique on the redbanded leafroller revealed that the insect uses three chemicals in its sex signalling. Now Dr. Roelofs wants to know exactly how the three-chemical signal works.

Do two of the chemicals say to the male, "You'll find me on apple tree X," and the third say, "Yes, indeed, I'm a female of your own kind and I recognize you as the right male"?

Ultimately, the goal is to find the combination of chemicals that would be most effective in disrupting the mating of leafrollers if manmade attractant were sprayed over the orchard.

Toward this end, chemist Jan Kochansky has prepared traps baited with different ratios of the three components, and entomologist Ring Cardé and Tom Baker are setting the traps out in the orchard to see which combination is most attractive to the insects.

The use of pheromones for pest control has many advantages over conventional pesticides, Dr. Roelofs said. "You only have to apply two to five grams of pheromone per acre, compared to pounds of insecticides; the pheromone can be applied with existing spray equipment; the pheromone compounds are very innocuous to all creatures except insects, and they are specific for one insect or group of insects."

Thus, pheromones could be used to control only the pest in question without disrupting populations of desirable insects and without harming any other form of animal life.

Actually, the redbanded leafroller emerged as a problem

only after DDT was widely used to control the codling moth, another costly orchard pest. The leafroller turned out to be relatively resistant to the effects of DDT, but its natural predators, which formerly kept the leafroller population in control, were not.

Now, the organophosphates sprayed every two weeks in the apple orchards are beginning to create similar problems with mites and aphids, Dr. Roelofs said.

A single pesticide may control several insect pests at once, but it also may destroy beneficial insects. On the other hand, Dr. Roelofs said, pheromone combinations might also be used to hit several pests at once, but without harming beneficial species.

Thus far, however, the elucidation of pheromone chemistry has outdistanced the development of the technology needed to apply this knowledge.

"If you just spray pheromone by itself, it's gone in a few days," Dr. Roelofs explained. "We need to enclose the pheromone in microscopic plastic capsules so that the chemicals evaporate slowly through the capsules over a long period of time—something like the microcapsules of the 'scratch and smell' books. That way the atmosphere can be continuously permeated with the odor of attractant."

11

TO BREED A PERFECT ELM

Bayard Webster

MILLBROOK, New York—Armed with a microscope, a warm greenhouse, millions of tiny elm seeds, and a lot of patience, a young scientist at the Cary Arboretum here is seeking the answer to a problem that may take him twenty years to solve—if he can ever solve it. Dr. David Karnosky, a twenty-seven-year-old forest genetecist on the arboretum staff, is trying to breed the perfect elm tree.

He wants to produce an elm that is beautiful and yet hardy enough to withstand the ravages of the Dutch elm disease, the fungus infection that has killed hundreds of thousands of American elms around the country and threatens the rest.

Dr. Karnosky would like to see a species that does not yet exist—a tree with the graceful crown and the larger leaves and the shade-giving shape of the American elm, but without its susceptibility to disease.

He would prefer an elm with some of the attributes of the Asian species, which are highly resistant to Dutch elm disease. But he would do without their smaller leaves and their tendency toward scraggly crowns because they would not provide the shade and the elegant ornamentation of the American elm, which has added its gentle grace to the streets and lawns of thousands of American towns. So it would seem that the logical thing to do would be to cross breed the different species until one got the combination of characteristics one was seeking. An elm, in other words, that could proudly wear many generations of robins' nests in its hair.

Unfortunately, however, trees cannot be crossbred the way different species of rabbits, mice, or hamsters can, which is to say, fairly easily and quickly. With these animals, conception to birth and sexual maturity takes from four to eight weeks. For an elm, it takes up to ten years to grow to sexual maturity—the point at which it can produce seeds.

But there are many problems in addition to time. The major one involves chromosomes—the microscopic strands of matter in the cells of all living things. These contain the genes that determine the characteristics of the plant or animal and its progeny. Chromosomes are usually constant in number for the same species and sometimes for the same genus.

To crossbreed different species of the same genus, or family, of plants or animals, usually requires that the two species have the same number of chromosomes. If the number is different, the so-called "chromosome barrier" prevents the successful propagation of a hybrid.

The problem with crossbreeding the American elm with the Asian elm is that the American tree has fifty-six chromosomes and the Asian only twenty-eight. But Dr. Karnosky and other plant-breeding researchers know that in some plants such as alfalfa, soybeans, wheat, and potatoes, twin seedlings (two plants growing from a single seed) occasionally produce a twin with half the number of chromosomes of the parent plant. Such knowledge has been used to change the characteristics of many kinds of agricultural crops.

Dr. Karnosky and others have reasoned that if American elm twin seedlings could be grown, some of them, like alfalfa and soybeans, might have half the number of their parents' chromosomes. This would be twenty-eight, the number of chromosomes in the Asian elms.

Then, if one seedling elm could be grown to maturity, it might be possible to cross it with an Asian variety. So far, according to Dr. Karnosky, no one has been able to grow an American elm that has been identified as having half the normal complement of chromosomes.

Aware that there were a lot of "ifs" in this kind of research, but undaunted and with the strong backing of the Cary Arboretum administration and staff, Dr. Karnosky set out on his long investigative road two years ago.

He began by collecting American elm seeds from some thirty trees in Dutchess County. He and his helpers first fanned out over the countryside in the spring of 1975, climbing trees, leaning from chimneys, standing on car roofs, clinging to "cherry-pickers," and picking the ripe seeds by hand and putting them into the pockets of carpenters' aprons they wore around their waists.

Then, after air-drying their harvest, Dr. Karnosky, like an indoor Johnny Appleseed, broadcast some 10,000 tiny elm seeds—each about half the size of a maple seed—on each of five beds of wet cheesecloth in the propagation greenhouse of the arboretum here, the 2,000-acre branch of the New York Botanical Garden. He had learned to do that at the University of Wisconsin while getting his Ph.D. in forest genetics in 1975.

Dr. Karnosky had found that the cheesecloth technique, with controlled light, heat, and humidity, produced one-inch-high seedlings in from seven to ten days with 90 percent germination. By dividing the twenty-foot-long seed tables into small grids, he could examine each plant to see if there were any twins. If there were, he collected those, discarded the remaining plants, and immediately scattered a new crop of seeds on fresh cheesecloth.

"We find about one twin for every thousand seeds," Dr. Karnosky said as he bent his head over the seed table. He noted that he and his assistants had examined almost 2 million seedlings since the project started.

So far, by a precisely controlled protocol, Dr. Karnosky has been able to grow 125 elm twins to heights of about twelve inches. Three of these twins have shown promise of having a reduced chromosome number. But it has not been shown conclusively that they have exactly half the normal chromosome complement.

To determine chromosome counts in the plants, the tips of their roots must be snipped off as they begin to grow through the hole in the bottoms of the red clay pots that are aligned, row upon row, in the greenhouse. This is because the root tips are the areas of the plants where most cell production is occurring and where the chances are best for finding a cell in a stage of division that most clearly reveals the chromosomes under a microscope. But before the tiny elms are strong enough to push their root tips through the bottoms of the pots, a series of delicate planting and potting operations has to be performed.

After collecting ten tiny twin seedlings that had been culled from some 10,000 green shoots on a single seed table, Dr. Karnosky carefully put them into a glass dish containing distilled water before taking them to his laboratory. In the laboratory, he lit a spirit lamp and briefly held a pair of tweezers and a scalpel in the flame. He then took ten small fibre planting pots from an autoclave where they had been sterilized. "The plant nutrients are right in the fibre," he explained.

Seated at a white table, Dr. Karnosky uncovered the glass dish and gently separated a set of twin plants with the tweezers, noting that one of the twins was smaller than the other. "We think that the small twins may be the ones with half the normal number of chromosomes," he said as he poked the smaller twin into the water-soaked solid fibre pot. He repeated the procedure with all the shoots and placed the fibre cubes in a plastic enclosure under fluorescent lights.

He next recorded the location and date of planting of each tiny green shoot, and gave each an identifying code number. Some of the tiny plants would die, he explained, pointing out that the smaller twins were also the weaker ones. About half of them, however, would slowly grow over several weeks until they could be transplanted into clay pots containing conventional potting soil. Then they would be moved from the laboratory to the more open environment of the arboretum greenhouse.

As he hunched over the microscope, looking for the tiny squiggles of chromosomes on a slide prepared several days before from a one-year-old elm root tip, Dr. Karnosky admitted that there were few auguries so far that his research would succeed in the end.

So far, he told a visitor, no one has been able to identify an American elm tree with half the normal number of chromosomes. "But there are quite a few scientists besides me who are trying to find one," he said. "It would be fun to be first," he added with a smile.

One problem he is facing is to find a staining technique that would more distinctly and clearly show and outline the bands and overall shapes of the individual elm chromosomes, which are only a few microns (thousandths of a millimeter) in length. If the types of chromosomes and their known characteristics in parent trees could be identified and then found in the supposed hybrid, he said, this could help confirm that hybridization had taken place.

To do this, Dr. Karnosky said, would also entail raising a twenty-eight-chromosome American elm to maturity, crossing it by distributing its pollen to the flowers of an Asian elm, collecting the seeds that were later produced by that elm, and then examining the characteristics of succeeding generations. All this could take up to twenty years.

To the young scientist, the road ahead looks long and

sometimes frustrating, but frequently exciting. In addition to his elm-breeding project, he is involved in testing a variety of trees for their resistance to urban air pollution. For this he uses a huge gas-chamber type of enclosure for the research in the arboretum's greenhouse. He will also do test plantings of young trees in New York City parks and along city streets.

In addition, Dr. Karnosky is monitoring the increasing acidity of rainfall in New York State in conjunction with the Boyce Thompson Plant Institute in Yonkers. And a twenty-year test of some 2,000 white ash trees collected from all parts of the country, to see how they survive in New York's climate, is just getting started on the arboretum's grounds under his guidance.

"I told my wife when I was a graduate student at Wisconsin that we'd have more time for recreation and seeing people when I got this job, but it's been just the opposite," he said. He and his wife and an Irish setter named Betsy live in a newly bought house in nearby Pleasant Valley.

"It's our first house," he said, announcing that he was going to spend the following weekend cleaning out the septic tank.

Out of the 125 twin seedlings he has nursed to the stage of young adolescence, one of them—"Number 33–2"—shows the most promise. "We think it may have twenty-eight chromosomes but we haven't been able to tell definitely yet," he said.

Looking up from the slide he had been examining, he said, "There's nothing here." It was a result he had encountered thousands of times before. But scientists live on hope.

"I think we're at the beginning point in forest genetics," he said, noting that with animals and some plants "you can turn generations over in weeks or months, but trees take years."

Back in the greenhouse, next to the pots of elms he had

grown, he noted that reality often impinged on hope. "If we can't justify this project in another year—that is, if we can't find elms with twenty-eight chromosomes—we can't justify the funding," he said ruefully.

When asked if there was not a better way of describing his project than to say it was like looking for a needle in a haystack, he thought a moment and replied with a smile:

"No. It's just like looking for a needle in a haystack."

12

A TOMATO THAT CAN STAND THE COLD

Jane E. Brody

GENEVA, New York—Snow had fallen and the temperature had dropped to 27 degrees that October night in 1972. So when plant breeder Dick Robinson went out to the field the next morning, he was more than a little surprised to see a few live tomato plants among the hundreds that had been killed by the frost.

This single serendipitous observation promises to give vegetable growers—both commercial and home gardeners—the first frost-tolerant tomato, which could mean a 50 percent longer season in which to harvest tomatoes in the nation's colder regions.

In talking casually to the modest breeder, he'd have you believe that this discovery—which could mean many additional dollars each year for New York's tomato crop—was only a matter of luck. "If we hadn't grown all those plants in that particular year, we would not have picked up the few that were frost-tolerant," he remarked the other day as he surveyed thousands of tomato seedlings in the greenhouse of the New York State Agricultural Experiment Station here.

When questioned further, however, Richard W. Robinson, professor of seed and vegetable science, revealed that in fact a lot of ingenuity, knowledge, teamwork, and toil helped to create the right atmosphere for luck to strike.

It began with Dr. Robinson's realization that "a radical

change" was needed in the tomato plant if New York and other northeastern states were going to compete with California-grown tomatoes.

In the last fifteen years, new varieties carrying disease resistance, earlier maturation of the fruit, and better-quality tomatoes had been developed for New York growing conditions by crossbreeding existing commercial varieties. But despite these developments, tomato acreage in the Northeast was falling and California was capturing an increasingly large share of the market here.

"We realized we had to go further afield than existing commercial varieties to find new genetic characteristics that might give us a better competitive advantage," Dr. Robinson said. So he began making crosses with wild species related to the tomato that grew in such places as the Andes Mountains and the Galapagos Islands.

"We turn to the wild species only as a last resort," he explained. "It would be a lot easier to work with domestic varieties, but when they don't have the characteristics we want, we look to what Mother Nature may have already selected out." One wild species, for example, has a natural protection against harmful insects—sticky hairs on the leaf surfaces act like flypaper.

However, the wild relatives of the tomato are inedible weeds, with fruits the size and color of a garden pea. And when crossed with the cultivated tomato, the wild traits dominate the next generation, called F-1, for first filial. "Nothing looks much like a tomato," Dr. Robinson said of the F-1 plants.

But hidden within the genetic mix of the tall, scraggly plants with tiny off-color fruits, Dr. Robinson found genes that would easily justify a twenty-year effort to breed them into commercial tomatoes. These genetic traits segregate out in the next generation, called the F-2, obtained by allowing the F-1 plants to self-pollinate. In the F-2 plants, genes from the original parents recombine in new ways, producing traits not previously expressed.

To make the crosses, the flowers of the cultivated tomato plant that will serve as the female parent must first be emasculated to prevent fertilization by the wrong pollen. Each pollen-bearing anther must be removed with a tweezer just before the flower petals open. Then the pollen is collected from the wild plant which will be the male parent.

For this job, Dr. Robinson and his assistants use an inexpensive handmade gadget—a converted flashlight, which serves as the power source, with the light replaced by the vibrating mechanism of a doorbell. Attached to the vibrator is a metal ring with a tiny plastic cup suspended beneath it. The ring is placed around the anthers of the flower, the flashlight-vibrator turned on, and the pollen shaken into the cup, ready to fertilize the emasculated flower.

"Out of 5,000 plants in the F-2 generation, no two looked alike, and not one was anything like a tomato," Dr. Robinson said. Most had no fruit at all; others grew tiny fruits of various colors; a few had fruits that matured earlier than either of the parents, and some showed resistance to certain insects.

"This illustrates what interests me most. You get completely new, totally unexpected types from gene recombination. Those little surprises—finding things the parents didn't have—are what make this work so fascinating," the forty-seven-year-old breeder said.

Though born and bred in the heart of Los Angeles, Dick Robinson discovered he was "really a country boy at heart" during World War II when he grew a Victory garden. "I saved some seeds from my vegetables and grew them the next year and got very different kinds of plants. They turned out to be hybrids produced by the bees, and I was intrigued by how different they could be from the original parents."

He continues to be intrigued by such things as orange tomatoes with lots of vitamin A that come from parents with red and green fruits but with little of the vitamin. He noted that Purdue University horticulturists have bred orange varieties with ten times more vitamin A than the typical tomato.

But it was the appearance of frost tolerance that most surprised and delighted the New York breeder. The original parents, the wild species and the domestic one, and their offspring, the F-1 hybrids, were all killed by frost that early October morning. But some of the plants in the F-2 generation survived.

Dr. Robinson harvested their seeds for further study in the station's controlled atmosphere growth chamber. Placing thermocouples on the plants, he found that at −3 degrees Celsius (27 degrees Fahrenheit) the parents and the F-1 hybrids were killed in two and a half hours, but the frost-tolerant progeny lived for up to eight hours. At lower temperatures, all the plants froze faster, but the frost-tolerant plants still lived much longer than expected.

"We don't hope to breed a really frost-hardy tomato, but rather one that could withstand a light frost of short duration such as we often have at night in April, May, September, and October. We've got only a six-week harvest season here. If we could extend the growing season three more weeks at either end, we could get a 50 percent increase in harvesting time," Dr. Robinson estimated. He pointed out that very often when the first tomato-killing frost comes, there are still a lot of unripe fruits on the plant.

Since 1972 when the frost-resistant plants first appeared, Dr. Robinson has been working to "purify" the genetic strain through repeated self-pollination, selecting to become the parents of the next generation those plants that are fertile and frost-tolerant.

"There was no hope, of course, that these crosses would look anything like a tomato," he observed. "Now that we have uniform frost-tolerance and good fertility, we're ready this season to cross back to the cultivated tomato, and I think we have a reasonable chance of getting some plants that will look like tomatoes."

He plans to release the experimental seeds this year to other breeders at agricultural stations and seed companies to speed up the ultimate development of a commercial frost-tolerant variety, about a decade from now. He anticipates it will take at least another five years of self-pollination and selection to breed out the undesirable characteristics of the wild parent and produce a plant that has the same desirable traits in every generation.

In addition to the problem of getting rid of unwanted genes, using wild species to improve cultivated varieties is complicated by the fact that many of the crosses fail to produce fruits with seeds for further breeding.

Dr. Robinson explained, "Because the two species are incompatible, the embryo sometimes aborts before the seeds form. So we have to remove the embryo soon after fertilization and grow it in a laboratory medium of agar gel and nutrients. There it develops into a small plant that can be grown in soil."

"Another trick," he added, "is simply to grow thousands and thousands of plants and hope that out of sheer numbers, you'll get one or two that will set fruit."

In making his original crosses, Dr. Robinson used seeds collected by Dr. C. M. Rick, his former genetics professor and mentor ("he made me fall in love with the tomato") at the University of California, Davis. "Some of the wild seeds Dr. Rick collected have a severe dormancy problem. They could lie in the soil for years and not germinate. Rick discovered that treating the seeds with laundry bleach would destroy enough of the seed coat to break dormancy."

As Dr. Robinson tells it, the California geneticist asked himself, "What could act as bleach in nature?" He guessed, "Digestive juices." Dr. Rick went to the Galapagos to assess the fate of the wild seeds after animals ate the fruits. He fed the fruits to Galapagos tortoises and studied their feces for days, but no seeds came through.

Mystified, he fed the fruits again, this time with a harmless dye as a marker. He waited and watched and finally, on the twenty-fourth day, out came the dye—and the seeds. When Dr. Rick planted them, the seeds germinated. Dormancy had been broken. He surmised that the lengthy time in the digestive tract would have permitted the animals to swim to another island, thus naturally dispersing the seeds.

But while Nature's main goal is dispersing a species, Dr. Robinson aims for developing a high-quality tomato ideally suited to New York growing conditions. Therefore, while selecting for such traits as size, shape, and color of the fruit, he will also pick for further breeding those plants that mature early and show resistance to diseases and insects.

Dr. Robinson is also breeding for better seed germination at low temperatures. Ordinarily, he explained, when soil temperatures are in the low 50s, tomato seeds don't germinate well, which prohibits starting them directly in the ground in spring.

Another factor he is working on is cold-tolerance for fruit set, or fertilization and development of the tomato embryo. If the temperature is below 50 degrees Fahrenheit (10 degrees Celsius) when the flowers appear, the fruit won't set because the pollen is damaged by the cold, Dr. Robinson explained. "This is why the first flowers on the tomato plant usually don't develop into tomatoes," he added.

Nor will cultivated tomato plants grow at temperatures below 50 degrees. So Dr. Robinson has crossed a commercial variety with a wild relative that grows luxuriantly in the cold 10,000 feet up in the Andes. The resulting hybrid grows well in the cold, but thus far all the plants have been sterile.

Tomatoes have yet another cold sensitivity that Dr. Robinson is seeking to eliminate—a failure to ripen to a normal red color if exposed to temperatures in the 40s or below. Until a cold-hardy type is developed, Dr. Robinson advises home gardeners to "pick their green tomatoes before the cold comes and ripen them indoors."

13
GENETIC DECODERS
Walter Sullivan

CAMBRIDGE, England—Using a new arsenal of chemical manipulations, scientists here and in several other laboratories are "reading" with increasing rapidity the genetic messages that control all life processes of an organism, from its origins until its death.

The research, in combination with new ways to alter these messages, has given birth to a revolutionary new field of biology. Its short-term goal is to understand how the message system works. Among its long-term goals are identifying the genetic bases of birth defects and cancer.

The messages are passed from generation to generation, encoded into very long molecules of DNA (deoxyribonucleic acid). A single such molecule can carry virtually all the instructions and information needed to control all vital functions of an organism.

In February 1977, the Laboratory of Molecular Biology here created a sensation when it published the full sequence of 5,375 "message units" in the DNA of the virus Phi X–174. A group in the laboratory is now more than halfway through the "reading" of the slightly longer DNA message of another virus, G–4. Both are "phages;" that is, they infect bacteria.

At Yale University, researchers, using somewhat different methods, have virtually completed deciphering a virus of comparable length known as SV–40 (for simian virus 40). It causes cancer in monkeys. An 800–link section of its genetic message seems peculiar to that virus in a manner suggesting a possible link to its cancer-causing properties. Once the new analytic methods have been applied to a number of viruses that have caused cancer and to others that do not, it should become clear whether those that do so share a common feature—a discovery of great potential importance.

A few years ago such penetrations into this innermost sanctum of life seemed hopelessly beyond reach. They have been made possible, to a considerable extent, by the chemical ingenuity of Dr. Frederick Sanger at the laboratory here. For nearly thirty years he has been devising ways to analyze long-chain molecules.

His first triumph, for which he received the Nobel Prize in Chemistry in 1958, was determining the sequence of sixteen amino acids that form the hormone insulin, whose deficiency is associated with diabetes. He then turned to RNA (ribonucleic acid), which distributes genetic information derived from DNA.

Most recently he has focused on DNA itself. It took nine researchers two years to determine the sequence of 5,375 units forming the DNA in the Phi X–174 virus. Now, using a much faster method developed by Dr. Sanger, two researchers—Dr. C. Nigel Godson of Yale, working with Dr. John C. Fiddes of the laboratory here—hope to finish the G–4 analysis in less than a year.

As evidence of the trans-Atlantic nature of this research, Dr. Fiddes will soon leave here to assist Dr. Howard M. Goodman at the University of California in San Francisco. Dr. Goodman, who earlier worked with Dr. Sanger here, reported recently that he had transferred from rats to bacteria the genetic code for synthesis of insulin.

Although Dr. Sanger rules the roost among the DNA analyzers here, he is very different from the two ebullient men, Drs. Francis H. C. Crick and James D. Watson, who first showed DNA to be a double helix—an interlocking braid of two molecular strands. That landmark achievement in the 1950s, vividly and breezily described in Watson's book, *The Double Helix,* took place here at Cambridge University, and in continuing the tradition of such research Dr. Sanger's quiet manner belies his extraordinary talent.

In the long journey from sequencing the sixteen amino acids forming insulin to DNA molecules with thousands of components one of his associates has been Bart G. Barrell, who began working here as a teenager. According to his colleagues he has an extraordinary memory for detail—even including sections of DNA code sequenced years ago. Because of his memory of past failures, successes, and analytical tricks over more than a dozen years of research here, he may, when a blind alley is encountered, suggest a procedure that almost worked a decade ago.

Such has been his contribution that, even though he lacks a college education, he is offering himself as a doctoral candidate at Cambridge University. His chances for success in that respect would be slim, according to his coworkers, were it not for the support of three Nobel laureates: Drs. Crick and Sanger and Dr. Max F. Perutz, overall director of the laboratory here.

The messages being deciphered are coded into chains of chemical units, or nucleotides, forming the long molecules of DNA. In all life forms, from the simplest virus to human beings, it is the sequence of these DNA nucleotides that determines the whole life process of that organism—its developmental timetable, its structure, function, and reproduction.

In the case of the virus Phi X–174, not only have the scientists here spelled out all 5,375 "letters" of the message, they have also, building on the work of other laboratories, made it possible to identify the roles played by most parts of the sequence. Thus the positions of segments that code for all nine proteins playing a role in the life of the virus are now known. Four of these proteins cluster symmetrically around the DNA core of the virus to form a coat with twenty facets. The five other proteins control replication and assembly of the virus as it proliferates inside a host bacterial cell and then breaks down the wall of that cell to liberate the virus particles. Both Phi X–174 and G–4 are phages that infect the intestinal bacteria known as *E. coli (Escherichia coli)*.

The most surprising discovery has been the way in which parts of the DNA chain can be read in two—and possibly three—ways. The life-controlling instructions coded into DNA are carried by only four "letters," or nucleotides, in contrast to the twenty-six letters used in English (with punctuation) to convey information. As in English, individual letters have no meaning. They must be grouped into "words." With DNA the words are all the same length, formed of three nucleotides. Some serve as punctuation, indicating, for example, the start and stop of a sequence coding for some substance or function. Others, in an extended sequence of "words," specify the building blocks (amino acids) needed to synthesize a certain protein.

What has been found, to the amazement of the scientific community, is that if the message is read, in terms of three-letter words, beginning with a certain letter, it has one meaning. But if the sequence is read, starting from the middle of a "word" in that message, completely different information is expressed. Dr. Fiddes likens this to discovering that, if you begin reading a book from the middle of a word, instead of at the start of a paragraph, what comes out is "a completely different story."

It is theoretically possible to read the DNA message three ways, beginning with the first, second, or third letter of a particular triplet. Current work hints that this may be the case at one or more points within the G–4 virus, whose DNA is slightly longer than that of Phi X–174.

Such an ingeniously contrived overlapping of genetic messages calls to mind the construction of a crossword puzzle. How it could have evolved naturally perplexes researchers in the field. It seems to occur, as well, in other viruses. According to Dr. Sherman M. Weissman, whose group at Yale has deciphered virtually all of the DNA in SV–40, there are overlapping sections within it, though fewer than in Phi X–174.

Whether this occurs in the DNA of higher animals, including man, is uncertain. Skeptics say that if this were the case, a change (mutation) in one gene could also affect a completely unrelated gene—a phenomenon not observed to date.

Analysis of the G–4 virus (short for Godson–4 in recognition of Dr. Godson's role in its discovery) is making it possible to lay its DNA sequence alongside that of Phi X–174, disclosing some features common to both and others that differ. For example, their start-stop functions seem different. Drs. Godson and Fiddes have found that the DNA of G–4, like that in Phi X–174, codes for nine proteins. Its biology, however, differs, for example, in the substances of its host cell that it exploits.

The differences in start-stop signals are of special interest, for normal functioning of life processes depends critically on their being turned on and off at suitable times. Dr. Sanger estimates that only about 5 percent of the DNA in higher organisms, including man, is used to produce proteins. The rest—much of it apparently repetitive —seems to control timing and structure. In human beings it causes each chemical reaction to occur when it should at every step of the embryonic, pubic, daily, menstrual, and reproductive cycles. In cancer the cellular control process seems to break down.

In the analysis of G–4 and Phi X–174 it has been

found that one "word," spelled ATG (for the nucleotides adenine, thymine, and guanine), seems to mark the start of a genetic signal, and that TAG often marks its end. Yet such "punctuation marks" may occur without playing this role. "Something rather subtle seems to be at work," says Dr. Sanger. If researchers can find out what it is, that will mark another milestone in understanding the life process.

Dr. Sanger is seeking another tenfold increase in the speed of DNA analysis and also hopes to develop a simple way for researchers to learn the sequence of links in a newly manipulated DNA molecule—the field of genetic engineering that has recently been a center of controversy.

The laboratory is operated within the Addenbrookes medical complex here by Britain's Medical Research Council.

Chemical Wizardry

The methods developed in the past year or two for rapid reading of the thousands of message units, or nucleotides, in the DNA "code of life" are masterpieces of chemical ingenuity.

Notable among them is the plus-and-minus technique devised at the Laboratory of Molecular Biology here by Dr. Frederick Sanger. A somewhat different method has been developed at Harvard University.

The life-controlling messages are coded into molecules of DNA (deoxyribonucleic acid) in terms of four "letters" or nucleotides of the DNA alphabet. These are A, G, C, and T (for adenine, guanine, cytosine, and thymine, the substances that characterize those DNA links). While the DNA of smaller viruses contains less than 6,000 such nucleotides, that of a human egg cell is thought to carry some 3 billion message units.

Normally DNA is formed of two strands of nucleotides twisted and interwoven to form a double helix. Each strand carries the genetic message, but one is complementary (rather than identical) to the other in a sort of chemical mirror image.

The nucleotides (A, G, C, and T) can be likened to four types of zipper teeth, each of which interlocks with only one of the other three. Thus, in binding together the two strands of DNA, A joins only with T. Likewise G mates only with C. If one strand carries the sequence A-G-C-T it can mate only with a strand whose sequence is complementary: T-C-G-A. They join thus:

```
A-G-C-T
||  ||  ||  ||
T-C-G-A
```

The molecule replicates itself by splitting into two strands, one of which, from a reservoir of free nucleotides, picks up a sequence of appropriate ones to reconstitute the other strand. The two DNA strands are popularly referred to as the "Watson strand" and the "Crick strand" for the two men who discovered their braided, interlocking structure.

DNA viruses under study here are single-stranded and circular, like a snake biting its tail, but when they invade a cell they rob nucleotides from the host to become double-stranded. This form multiplies until there are about twenty of them. These then begin producing new single-stranded virus particles coated with protein. After some 200 have been generated, the computerlike DNA control system produces a protein that breaks down the wall of the host cell, allowing the new virus particles to escape and infect other bacteria.

The plus-and-minus analytical method uses, as its starting material, the double-stranded form. The analytical steps are as follows:

1. STRAND SEPARATION

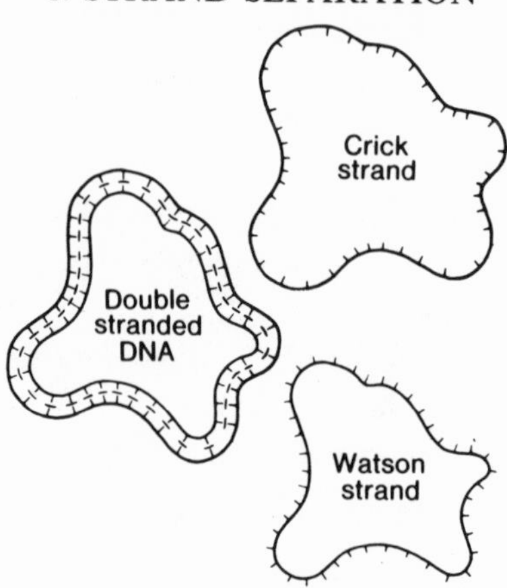

Some double-stranded DNA is treated chemically to split apart the Watson and Crick strands.

2. CLEAVAGE

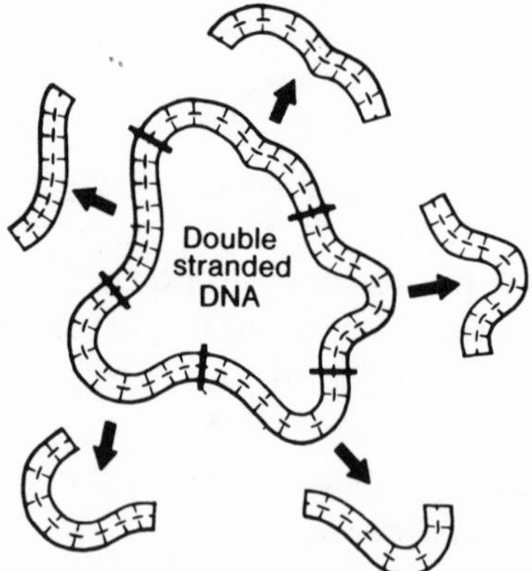

Other samples of double-stranded DNA are cut into segments by "restriction enzymes" known to break specific bonds. There is a sufficient variety of these enzymes so the

DNA can be cut up many ways for successive analyses.

The segments are separated by electrophoresis, a process in which particles suspended in a gel migrate in response to an electric field. Their migration rates depend on their size. When the current is turned off, each type of segment has reached a separate region of the gel. These are cut out and the segments extracted to be used, one by one, to "prime" a DNA strand for analysis.

3. ANNEALING

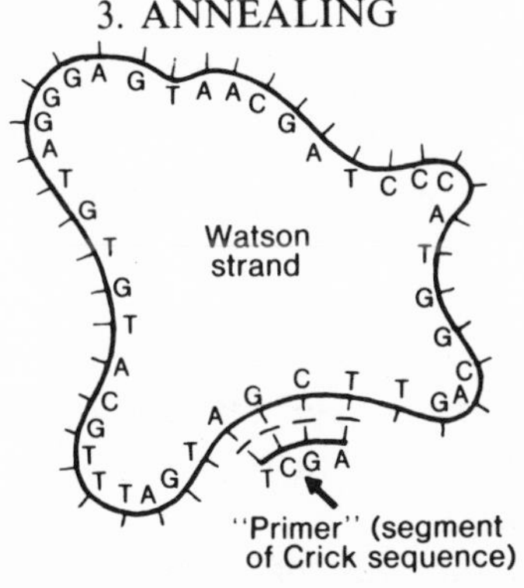

The segment selected as "primer" for a particular analysis is "annealed" to strands of DNA (from step 1) by incubating both substances in solution. If the DNA strands are of the Watson variety, the Crick side of the segments will seek out and attach themselves to the matching sequence of the Watson version. The other (Watson) half of the segment will be cast off.

4. ADDING ONTO THE PRIMER

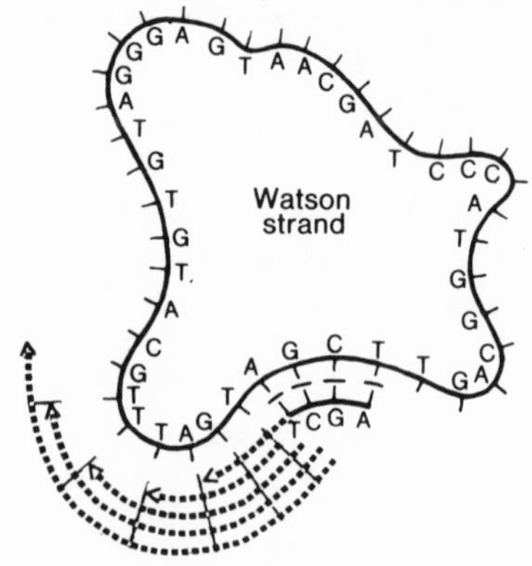

The preparation is then incubated with a mixture of all four nucleotides of the DNA alphabet (A, G, C, and T), at least one of which is tagged with radioactive phosphorus, plus an enzyme (DNA polymerase) that stimulates growth from one end of the primer. Starting there the appropriate nucleotides begin attaching themselves, reconstructing the broken DNA chain.

Thus if, as in the illustration, a T lies opposite the first vacant site at the end of the primer, an A will attach itself, followed by a C to match the G on the Watson strand, and so on. After a time ranging from a minute to a half hour the process is halted chemically. The time interval is chosen so the growing strands will have reached the section targeted for analysis.

The strands do not all grow at the same rate and, therefore, when their growth is stopped there is a complete sampling of chain lengths for the area to be analyzed. If that area, for example, extends from positions 110 to 260 along the DNA chain, some chains end at each position in that series.

Dotted lines in the diagram indicate molecules that have

reached various stages of growth from the primer when growth is stopped.

The sample is now split into eight subsamples, or aliquots, for the final stage of the analysis.

5. THE MINUS METHOD

Each of four aliquots is incubated with a mixture of DNA nucleotides in which one of the four is missing, plus a growth-stimulating enzyme. The strands whose growth was halted in the previous step begin growing again. This time their growth halts when it reaches a point where the missing nucleotide should be added.

That is, if the test sample is lacking the A nucleotide, its growth will stop where there is a T on the matching (Watson) strand, since it will only pair with an A. Because the growing strands at the start of this stage represented a complete sampling of lengths, they will grow to a succession of points where A nucleotides are required. The result is a piling up or clustering of molecules just short of each such point, like cars at a succession of red traffic lights. Because the nucleotides have been tagged with radioactive phosphorus, it is now possible to determine where in the DNA sequence such clustering has occurred.

The restriction enzyme that was used to cut out the primer segment now lops it off, leaving the newly formed chains for analysis. Again electrophoresis is used, exposing the sample to an electric field.

The shortest chains migrate down the column of gel most rapidly. The longest ones linger behind. After about three hours the chains have spread into a succession of bands, like the rungs of a ladder. These cannot be seen until they have spent the night against a sheet of x-ray film.

When developed, this shows where the radioactive material ended its migrations, the dark rungs being where an absence of the missing nucleotide caused a blockage.

6. THE PLUS METHOD

This step, in a sense, reverses the minus method. Each of the four remaining aliquots, or subsamples (in tiny glass vials) is exposed to an enzyme that begins eating away the chains of varying lengths that grew in step 4. However, in each aliquot one of the four nucleotides is provided in sufficient abundance to prevent the enzyme from chewing off that particular link in the genetic chain. Thus, each chain is cut back until that nucleotide is reached.

As in the minus method there are clusterings of chain lengths at such points whose locations can be shown by electrophoresis and x-ray film exposure. The analysis is rapid because all eight aliquots can be processed side by side, producing eight ladders that reveal, by each method, locations of all four nucleotides. Thus, a hundred or more links at a time, it is possible to read DNA messages thousands of links long.

14

A STRANGE BACTERIUM'S PURPLE PIGMENT

Jane E. Brody

SAN FRANCISCO—From the air it looks as if someone dumped red food coloring into the salt flats at the southern tip of San Francisco Bay. In fact, the red-orange hue of the drying salt is due to vast colonies of bacteria that seem to defy the laws of life: they can survive without special protection in a dry salt crystal.

Salt ordinarily kills bacteria and therefore has been used since the days of early man to preserve foods. But these salt-loving bacteria grow best when the salt concentration is seven times that of sea water. They actually disintegrate at low salt levels.

To the salt industry here and elsewhere, these organisms —called halobacteria—are one large headache. It is necessary to heat the dried salt to 800 degrees Centigrade to burn them off.

But to Dr. Walther Stoeckenius, biochemist at the University of California at San Francisco, and to a dozen or so other researchers in this country and in Europe, the salt-loving bacteria are a challenging biological, chemical, and physical puzzle that can provide new insights into basic life processes.

Although only microscopic in size (about 5,000 bacteria lined up end to end would equal an inch), this rod-shaped cell houses in its membrane a gold mine of biochemical information that may ultimately find applications in medicine, agriculture, solar energy, and the search for life on other planets.

For, as Dr. Stoeckenius and his colleagues recently discovered, beneath the organism's red-orange "mask" lies an extraordinary purple pigment that can convert sunlight into electrochemical energy. A close relative of the visual pigment in the eyes of man and other animals, the purple pigment in halobacterium's membrane is the only known biological substance besides the chlorophyll of green plants that can change solar energy into other useful forms of energy.

Whether the bacterium or its pigment can ever be harnessed to produce significant amounts of power from sunlight is open to question. But research into the fine workings of this single-cell powerhouse may provide a better understanding of certain muscle and nerve diseases, metabolic defects that result in obesity or extreme weight loss, visual disorders like colorblindness, and even possibly the nature of cancer cells.

The purple pigment may also lead to an economical new approach to desalinization of water and to the construction of a biological solar cell, perhaps for use on long space flights. If nothing else, the purple membrane bacterium may provide clues to life elsewhere in the universe where conditions are hostile to most forms of life on earth.

Yet for all it may do or reveal, the purple membrane is an extraordinarily simple system of generating energy that can be turned on or off merely by switching the light.

The first hint of what biochemical excitement lay beneath the bacterium's red-orange coat grew out of an erroneous scientific report. In 1964, Dr. Stoeckenius, a German-born student of cell membranes who was then at Rockefeller University, came across a report stating that halobacteria had no cell walls and had membrane characteristics which ran counter to his own then-unpopular theories about membranes.

In investigating the claims, Dr. Stoeckenius, working with Robert Rowen, found cell walls and more support for his own membrane theories. But in the process, what he discovered about halobacterium's purple membrane changed the entire direction of his research. Instead of devoting his studies to the structure of various cell membranes, Dr. Stoeckenius has for the last thirteen years pursued the photobiological puzzle of the purple membrane bacterium.

"This was simply too good to leave for someone else to investigate," Dr. Stoeckenius remarked over lunch recently, his bright blue eyes sparkling with the recalled excitement of his early discoveries about halobacteria.

With his colleagues at Rockefeller University, Dr. Stoeckenius broke up the bacterial cells and whirled the parts in a centrifuge, which sorted them according to their size and buoyancy.

Three main fractions resulted: an unsurprising red-orange portion colored by carotene pigments, which protect the bacteria from damage by blue light in their normally sunny environment; a yellow segment of gas-filled sacs that probably help the bacteria to float at a certain depth, and a mysterious deep purple fraction rich in protein.

The scientists made micrographs of the purple material through an electron microscope. By relating what they saw to how the cell membrane diffracted x-rays, they deduced that the purple stuff was present in discrete patches in the membrane. But they had no idea what purpose it served.

The search for an answer took Dr. Stoeckenius to San Francisco where, under a National Institutes of Health project grant to the University of California's Cardiovascular Research Unit administered under "an enlightened director," he was given the money, space, and freedom to pursue whatever research question challenged him. Such "open" grants, he noted, no longer exist; today each project must be approved separately by the institutes.

In 1969, under Dr. Stoeckenius's open grant in San Francisco, the exploration of the deep purple mystery was undertaken by Dr. Allen E. Blaurock and Dr. Dieter Oesterhelt, a scientist from Germany. X-ray diffraction studies by Dr. Blaurock and chemical analysis by Dr. Oesterhelt revealed a single protein containing a lot of water-avoiding amino acids arranged with fatty material in a hexagonal structure.

But the purple pigment's physical composition told little about why it was there or how it may work. So next the scientists examined its possible functional characteristics. They analyzed the wave lengths of light that the purple material absorbed, finding a peak between 560 and 570 nanometers in the green part of the visible spectrum and another at 280 in the ultraviolet part, which is characteristic of proteins. But when they tried to isolate the factor responsible for the light-absorbing pattern, the purple color just "disappeared" whenever they added a solvent or detergent to extract it.

Then came that remarkable process of memory-jogging and mental synthesis that led Dr. Stoeckenius and his colleagues to the ultimate answer. Maybe, they guessed, the purple pigment was something like rhodopsin, the "visual purple" that helps man and other animals to see.

Indeed, biophysical analyses of the two pigments showed a remarkable similarity, suggesting that the purple membrane pigment had light-absorbing properties that were important to the functions of the bacterial cell. "Bacteriorhodopsin," as Dr. Stoeckenius dubbed the bacterial counterpart of rhodopsin, was shown to undergo reversible bleaching when exposed to light, indicating that light changed it chemically.

The California scientist told Dr. Oesterhelt, who by then was back in Germany, of this discovery, and in his specially equipped laboratory in Munich, Dr. Oesterhelt found that when bacteriorhodopsin underwent reversible bleaching, something else also happened—positively charged hydrogen ions, or protons, moved into and out of the purple membrane, suggesting that the membrane pigment might function to produce energy for the bacterial cell.

Meanwhile, Dr. Stoeckenius, working with scientists at two other American universities, explored the workings of the purple membrane from another angle. They discovered that exposure to light "drove" the pigment through a repeating cycle of absorption changes, accompanied by the continual release and uptake of protons.

As the California scientist recalled the way the mystery was unraveled, "The first clue came with our observation that in the light the purple membrane bacteria synthesized ATP," the main energy storage molecule in cells. Usually, he explained, respiration in the cell and the synthesis of ATP happen together. But when cells are treated with substances called "uncouplers," ATP synthesis and respiration are no longer linked. Instead, the cells stop ATP synthesis but increase their respiration.

When an uncoupler was used on halobacteria, the light-stimulated synthesis of ATP stopped. Since uncouplers are known to facilitate the passage of protons through cell membranes, "we reasoned that light caused the cells to eject protons," Dr. Stoeckenius said. "So we looked for protons and we found them."

He took the problem to Dr. Ephraim Racker, a biochemist at the New York State College of Agriculture and Life Sciences at Cornell University and an expert on proton transport. Dr. Racker, whose white lab coat is perennially stained purple from his work with bacteriorhodopsin, made artificial capsules called liposomes, incorporating the purple pigment and the enzyme that makes ATP into the liposome shell. The liposomes were suspended in a medium containing ADP and inorganic phosphate, the building blocks of ATP. When light was shone on the liposomes, protons were pumped and ATP was made inside them.

Dr. Racker was amazed: "Bacteriorhodopsin, a single polypeptide, can replace the entire respiratory chain of fifteen polypeptides that is ordinarily needed to make ATP. This is the simplest known system to create biological energy in the form of ATP, the universal energy currency in living cells."

Bacteriorhodopsin, then, functions as a "proton pump," generating energy each time it ejects a proton. It does this by converting light energy into an electrochemical gradient—a different concentration of hydrogen ions on one side of the membrane than on the other. The energy stored in such a gradient is like water behind a dam; it can be translated into biological work or into electrical energy.

At the Ames Research Center of the National Aeronautics and Space Administration in California, researchers showed that each molecule of bacteriorhodopsin can pump 200 protons per second across the cell membrane. Dr. Stoeckenius said work in several laboratories has shown that a single cell containing purple membrane can create an electrical potential of 0.3 volt, enough to distill water.

Dr. Russell E. MacDonald, a Cornell biochemist, and Dr. Janos Lanyi, a biophysicist at Ames, have found that halobacterium's proton pump can be used to transport essential nutrients into the cell and pump sodium ions out. Therefore, Dr. MacDonald said, one possible use of the pump is to create a light-activated desalinization system. Areas like southern California and Israel, where there is a lot of sun and little fresh water, could make good use of such a system, he said. Thus, whereas the salt industry works hard to get rid of halobacteria, the desalinization industry might use the bacteria to get rid of salt.

NASA is interested in the possibility of using purple membranes to create large biological "batteries" that could operate continuously on solar energy and perhaps provide power on long space flights. Dr. Stoeckenius said that the Russians also have a large research group working on a solar cell based on purple membrane.

Dr. Racker said he works with the purple membrane "almost daily" as a tool to explore biochemical interactions that may be pertinent to various human maladies. For example, the white-haired scientist said, "uncouplers" can cause animal cells to lose energy and have been used to treat obesity. At times, however, uncouplers work too well, killing the patient while ridding him of excess poundage. Dr. Racker is using bacteriorhodopsin in liposomes to test how uncouplers work, with the idea that safer ones might eventually be designed.

In another application, Dr. Stoeckenius said, the extremely simple proton pump of the purple membrane may help in understanding more complex ion pumps—such as the movement of sodium, potassium, and calcium in and out of cells—which are important in muscle contraction and nerve conduction and which may go awry in certain neuromuscular diseases. The bacterial proton pump may also help to elucidate the pumping of amino acids to nourish cancer cells, he added.

And the National Institutes of Health, which helps to support bacteriorhodopsin research, is interested in using the bacterial pigment as a tool to unravel the molecular workings of man's visual pigments. The stimulation by light of rhodopsin in the eye is the first step in the process of visual perception, Dr. Stoeckenius explained.

He added that studies of visual pigments and how the proton pump works on a molecular level will be greatly aided by the discoveries of two molecular biologists from Cambridge, England—Nigel Unwin and Richard Henderson. Using a series of tricks to enable them to see extremely fine details under the electron microscope that were previously too small to be distinguished, the Cambridge scientists worked out the internal structure of the purple membrane. They were then able to construct a three-dimensional model of the purple membrane protein.

According to Dr. MacDonald, this made bacteriorhodopsin "the first membrane protein whose three-dimensional structure is known and the first membrane molecule for which both the physical structure and function have been deciphered."

15

UNCOVERING SECRETS OF INSECT LIFE

Boyce Rensberger

NAIROBI, Kenya—Among the towering eucalyptus trees on the Masonga Wai River here stands one of the most remarkable, if least-known, scientific institutions in the world—a complex of laboratories where more than thirty researchers from seventeen countries are discovering the innermost secrets of the lives of insects and closely related species.

They have found what makes wood taste good to termites; how to induce abortions in tsetse flies, one of the few insects that incubates its eggs in a uterus; how ticks in the grass find other ticks already parasitizing an animal and join them in clusters, and how termites decide when to build a roof over their trails.

The scientists say such knowledge is essential to developing effective and environmentally sound pest-control methods. They also hope to use the same knowledge not to destroy insects but to grow them more abundantly under controlled circumstances as sources of food and other products.

Established in 1970, the International Centre of Insect Physiology and Ecology is regarded throughout the scientific world as the leading institution of its kind.

I.C.I.P.E., pronounced every way you can think of by its polyglot staff, is an independent, internationally supported center that concentrates on seven types of insects and arachnids responsible for enormous crop damage, livestock destruction, and human suffering, particularly in tropical regions. The target species are tsetse fly, termite, mosquito, tick, African army worm, maize stem borer, and sorghum shootfly.

"These are extremely difficult pests which have defied solutions for decades. Ordinary pest-control techniques have not been successful and, out of necessity, we have had to reexamine these insects from the most fundamental basis," said Professor Thomas R. Odhiambo, the center's director.

Prof. Odhiambo, a Kenyan and an internationally recognized authority on insect reproductive hormones, conceived the idea for the center ten years ago as both a much-needed center for attacking difficult insect pest problems and as a focus of scientific excellence that he believed developing countries needed to become truly independent.

Prof. Odhiambo proposed an insect biology center in a 1967 article in *Science* magazine entitled "East Africa: Science for Development." In that article he argued that, unlike Western philosophies which are dualistic, distinguishing between objective and subjective realities, traditional African philosophies are monistic, making no such distinction.

"As a result," Prof. Odhiambo wrote, "science in the modern sense has no firm foundations in African society." He argued that newly independent African countries would remain handicapped if they only imported the technological fruits of science done elsewhere and failed to develop their own centers of scientific innovation.

Such centers in African countries, Prof. Odhiambo asserted, should not only tackle African problems but also, by exposing more Africans to scientific ways of thinking, serve to revise traditional outlooks on life that hamper economic and technological development.

Prof. Odhiambo's article caught the attention of a number

of Western scientists who were also interested in the Third World. Led by Carl Djerassi, a Stanford University biochemist, they agreed to work with Prof. Odhiambo to turn his idea into reality. Their success has been nothing short of remarkable and, by most accounts, largely the result of Prof. Odhiambo's tireless efforts to persuade private and governmental bodies in a number of countries to donate money to support the center.

Among the center's prime targets is the tsetse fly, an insect resembling a horsefly that infests some four million square miles of Africa, and carries African sleeping sickness, also called trypanosomiasis. The disease accounts for substantial human suffering and death and puts vast tracts of prime grazing land off-limits to cattle.

Whereas most insects reproduce prolifically, laying enough eggs to sustain massive losses and still keep the species going, tsetse flies produce only one offspring at a time. Instead of laying eggs that immediately become vulnerable, the egg is carried in a uterus where it hatches and the larva fastens its mouth on an intrauterine "milk" gland and matures. Just before it is to pupate, a cocoonlike phase, the fully mature larva is born. After pupation, larvae become adult flies. Because of this maternal protection, the larvae are not readily vulnerable to pesticides.

"But this complex reproductive system gives us an opportunity to look for some weak link that we can exploit," said Dr. Mohammed Chaudhury, an American scientist at I.C.I.P.E. He and his colleagues have identified a number of hormones produced by various glands in the mother tsetse that control the reproductive process.

By performing surgery on individual flies to remove the glands, Dr. Chaudhury is finding out which hormones control which steps of the process. Using a binocular microscope and instruments designed for eye surgery, Dr. Chaudhury carefully opens the fly, which is chilled in cold water to immobilize it, and excises the tiny organ. The wound closes without stitches.

"At first we had a lot of mortality," Dr. Chaudhury said, "but eventually we improved the technique and haven't lost very many flies."

He found which gland and hormone trigger ovulation. Removal of the gland rendered the flies sterile. He found another hormone that controlled the milk gland. Removal of this one caused the next two or three flies to be born as miniatures, and the one after these to be aborted before term. Further reproduction was blocked entirely.

Tsetse researchers at I.C.I.P.E. are also testing the effects of overdoses of hormones. Some have been synthesized artificially and administered to the flies to produce abortions. Such a chemical would be useful in the field for controlling tsetse flies if a way could be found to deliver the hormone in the right doses to the flies. Dr. Chaudhury's colleagues are working on one novel approach: feeding the hormone to the cattle on which the flies feed. The hormone should be harmless to the cattle but flies that sucked their blood would ingest a dose and be unable to reproduce.

A similar strategem of armoring cattle against their parasites is being investigated for tick control by Dr. Fred Obenchain, another American at I.C.I.P.E. Ticks are a major threat to cattle in Africa because their bites transmit a fatal disease known as East Coast fever. Even ticks that do not carry the disease may attack by the thousands, each taking one to two centimeters of blood, causing the animal to weaken and die.

"Immunizing the animal against the disease still leaves it vulnerable to sheer blood loss from nice clean ticks," Dr. Obenchain said. "We're going to try to immunize animals against the ticks themselves."

The goal is to develop a vaccine from some critical protein substance within the tick, perhaps the hormone necessary for ticks to molt periodically. After such an immunization, cattle would carry antibodies to the hormone in their blood. When a tick took its blood meal, it would ingest the antibodies which, in turn, would destroy the molting hormone in the tick. Without this hormone the tick could not shed its old exoskeleton and grow. It would die before it could reproduce.

Termites are another of the center's major targets, not only because of their propensity for eating wood in housing but because several species, especially in Africa, eat grass, competing with livestock and wildlife.

Dr. Gilbert W. Oloo, a Kenyan, has been studying termite food preferences and how the insects find their favorite foods. He has made chemical extracts from various grass species and identified the substances that contribute the preferred flavor. Dr. Oloo once soaked some filter paper in a flavor extract and presented it to the termites in one arm of a Y-shaped maze while their favorite grass, depleted of its flavor, was in the other arm. The termites preferred the paper.

Dr. Oloo and his colleagues study termites in the field, where they burrow in the ground like ants, but the scientists also keep a number of colonies in the lab, each in a block of earth carefully dug out of the ground.

Dr. Oebele H. Bruinsma, a Dutch termite specialist, is analyzing the ways termites build their nests, mounds, and foraging trails. The behavior is often governed by chemical secretions. For example, the large body of the queen termite has been found to emit a volatile substance. When worker termites smell this substance at a certain concentration, they gather soil particles in their mouths and begin cementing them together at precisely the distance from the queen's body where the odor is at the right concentration. As the wall gets higher, the concentration of the odor diminishes, so the workers begin making the walls curve inward until they meet, forming an oval, queen-shaped dome known as the royal cell.

A similar phenomenon prompts termites to build roofs over regularly used foraging trails. When a new trail is established, it simply runs over open ground. As the concentration of odiferous substances inadvertently deposited by the traveling termites grows, it reaches a critical point that suddenly stimulates the termites to gather soil particles and cement them together, forming a roof over the trail.

Dr. Oloo has obtained quantities of the substance, painted squiggly lines across pieces of paper, and watched the termites dutifully roof over the circuitous paths. Such experiments are more than scientific tricks. If termites depend so heavily on various odors to survive, interfering with these chemical messages through some nontoxic process may prove a better method of termite control than hazardous pesticides.

"This is really a major aspect of our program," Prof. Odhiambo said. "The old methods of using chemical pesticides have been found to have disadvantages. Biological controls can give us safer and more effective methods of pest management."

Prof. Odhiambo has another idea. Once some of the problems of insect pests have been dealt with, he would like to turn the same intimate knowledge of insect life toward raising insects as food. In many parts of Africa, various insects are eaten as food, not only by people on the edge of starvation but also by many who consider termites, for example, to be a tasty part of a varied diet.

Prof. Odhiambo believes that it may be possible to farm insects on an intensive basis and produce large quantities of protein in addition to other insect byproducts such as honey and beeswax.

"I would like us to have a shot at it," he said, "but funding is still a tremendous problem. It has taken a long time to win donors." Prof. Odhiambo now must spend much of his time visiting donor groups around the world to maintain and increase the center's budget. It has grown over the years but still lags behind what the center's officials feel it should be.

"One mission of I.C.I.P.E. is to help establish science in East Africa but, as a nongovernmental institution, we need outside help to strengthen our base," Prof. Odhiambo said. "Developing countries must develop the capability of doing science. It is a necessary part of being independent."

16

A JUNGLE FIELD STATION

Walter Sullivan

BARRO COLORADO ISLAND, Canal Zone—The rain forests of Panama are a reservoir of malaria—but no longer the kind that infects human beings and thwarted French attempts to build a canal here. That variety and its fellow scourge—yellow fever—have been largely eradicated by mosquito control. For reasons still stubbornly mysterious, the type of malaria that virtually saturates the forest infects lizards.

Knowing how it is transmitted and why it does not seem to have a dramatic effect on the lizards would fill an important gap in the understanding of parasitic disease. Stella Guerrero and her mentor, Dr. A. Stanley Rand, are engaged in a search for the answers—for example, the as yet unidentified mode of transmission.

Their colleagues at the field station here are investigating other problems: whether certain giant trees "commit suicide" after fruiting to make room for their offspring in the dense jungle, whether other trees of the region can grow only with the help of root fungi, and why certain caterpillars inhabit unmolested the nests of vicious spiders.

The field station, which can accommodate up to thirty-eight researchers, is a branch of the Smithsonian Institution's Tropical Research Institute. It was set up after Gatun Lake was flooded to form the central part of the Panama Canal, creating this 6,000-acre island almost entirely buried in lush tropical forest. The island was made a biological reserve in 1923.

Dr. Rand, who is assistant director of the institute, has also been studying iguanas, larger cousins of the tiny malarial lizards, which are only a few inches long. He has found that, contrary to assumptions regarding reptiles, they behave in a social manner. Follow-the-leader footprints left by prehistoric dinosaurs indicate that they, too, displayed group behavior.

The malaria infections, Miss Guerrero and Dr. Rand suspect, may explain why the lizards seem to underpopulate their environment. So far, however, they have been unable to determine just how the disease affects the reptiles.

In malaria-free environments reptilian populations often seem to be limited only by the availability of food. Here and in other areas where lizard malaria is endemic, as in Puerto Rico, food is ample and some other factor, such as predation or disease, seems to limit population growth. In fact the researchers have found that, despite the highly stable climate, the lizard population of Barro Colorado may vary tenfold from year to year.

Those under study are a tiny species of the genus known as *Anolis,* from the Haitian Creole name for them (anole). They are trapped once a month in a study plot twenty by fifty meters (a meter being slightly more than a yard). From thirty to fifty lizards are caught each time, depending in part on seasonal population variations. It is estimated that the total population of the plot ranges from 50 to 150. Miss Guerrero gives each captured lizard a medical examination, including microscopic study of its blood to see if the cells are infected by malarial parasites. The lizard is tagged and its condition catalogued before it is released at the plot.

During the past year a number of the tagged lizards have been recaptured; a running case history is maintained for each. So far, no evidence has been found that the infection has adverse effects. Some animals are now being studied more intensely in captivity, fed on termites, roaches, and fruit flies, as well as crickets flown in from Louisiana. Lizards carrying the disease seem to eat as much, grow as fast, and live as long as the others. This leaves impaired reproduction as the most likely adverse effect and an effort is being made to test that hypothesis.

While the most virulent form of human malaria is caused by the parasite *Plasmodium falciparum,* carried from person to person by the *Anopheles* mosquito, no clue has been found to the transporter, or "vector," of the disease so far as lizards are concerned. Birds, such as Canada geese, are infected by one of the Culex group of mosquitoes, but according to Dr. Rand a researcher in Georgia, seeking the vector for lizards, has examined every candidate evident in that region without finding such a carrier. Attempts are being made here to produce infections in the laboratory.

The lizard disease is caused by *Plasmodium balli* or *Plasmodium tropidari* (which some specialists divide into two species). None infect human beings. One finding from Miss Guerrero's blood studies has been that some of the captive lizards seem to shed the disease and a percentage of these later contract it again—possibly because it had been dormant. Reptilian blood cells, in contrast to those of mammals, have nuclei and live longer, which Dr. Rand believes may be a factor.

After infection, parasites multiply in the blood stream, whereupon the lizard—in a manner not yet identified—sheds a large percentage of its blood cells, including infected ones. For a time the animal is clearly anemic, but it then settles into a state of mild anemia and mild malarial infection.

Miss Guerrero, a native of Colombia, worked here for three months in 1974, then returned to the Universidad del Valle in Cali, Colombia, and earned a bachelor of science degree. In April 1976, she resumed research here under a grant from the Exxon Foundation. When it expires she hopes to begin graduate work in the United States. The grant is part of an Exxon program to assist students from Latin American countries.

The research station is equipped with air-conditioned laboratories, communal eating facilities, and tin roofs that periodically resound with the scampering of local bands of spider monkeys. The night air is filled with the lionlike roars of howler monkeys.

Dr. Rand's iguana research has been conducted with Dr. George M. Burghardt and Dr. Harry W. Greene of the University of Tennessee. Because iguanas like to lay their eggs underground in a clearing, and such clearings here are rare, researchers have found that as many as 150 to 200 females use a single plot measuring six by seven meters.

When the eggs hatch the babies begin poking their heads out of holes, apparently looking around to see if others are ready to venture into the world. Such behavior often continues an hour or more. Sometimes eight heads squeeze out of the same hole. Rarely do the reptiles emerge and move off alone. Often they do so in single file, in the manner inferred from dinosaur tracks.

Since 1974 Dr. Rand and his coworkers have observed juvenile iguanas for more than 500 hours. In a report published in *Science* they say their findings "indicate sophisticated behavioral mechanisms we believe to be previously unrecorded in reptiles, and they suggest changes in our thinking on the evolutionary origins of vertebrate social behavior."

Such group activity was not seen in turtles or crocodiles nesting in the same area. Baby crocodiles peep when hatched and are dug out by a parent. Almost all the iguanas swim to other islands in groups, attacked at times by birds or other reptiles. Most survive by diving or running across the water. Dr. Rand and his colleagues believe that, by migrating together, more iguanas survive than would otherwise be the case because local predators can attack only a limited number of them during one such migration.

Caterpillars that inhabit spider webs have been studied here by Dr. Michael H. Robinson and his wife Barbara. Webs spun by colonies of the spider *Anelosimus eximius* are scavenged by caterpillars feeding on the solid remains of insects whose other parts have been liquified by digestive juices of the spiders before ingestion. Such spiders are habitually fierce, attacking anything that moves (or smells foreign?) in their webs. The caterpillars live there with impunity—possibly, the Robinsons believe, because they hatched inside the web and have acquired an odor that identifies them as natives.

An important role of the station is environmental monitoring as part of the worldwide effort to identify adverse trends. The towering forest provides a variety of habitats, from the leaf-littered ground to the highest canopy of branches. A 138–foot tower enables researchers to make observations and collect insects and other samples from ground level to the treetops.

Some 400 trees, representing 100 species, are checked weekly to monitor the timetables of their flowering, fruiting, and leaf shedding. By keeping detailed records of weather, ground moisture, and water runoff, researchers hope to identify factors that determine the health of such forests, which occur—and are being threatened—in many tropical regions.

17

BEARS: A SEARCH FOR THE SLEEP HORMONE

Lawrence K. Altman

ROCHESTER, Minnesota—In open-air cages on a farm at the Mayo Clinic here, two black bears have gained more than 100 pounds each after feasting around-the-clock for several weeks preparing to go into a long winter sleep.

The bears are about to be moved into dens where, from time to time, a team of Mayo scientists will join them for stretches lasting as long as thirty-six hours and in temperatures that drop to as low as −30 degrees Fahrenheit. The scientists will be drawing samples of bear blood.

For several winters, Dr. Ralph A. Nelson's team here has tried to unlock one of nature's greatest secrets—how, even when amply supplied with fat, the bear can sleep for up to five months, burning 4,000 calories each day, and yet not once eat, drink, urinate, or defecate.

The scientists are seeking a hormone that they suspect controls the bear's winter sleep pattern. Discovery of such a hormone, they believe, might offer new ways to treat human diseases such as kidney failure, sleep disorders, obesity, starvation, and open new avenues of research into many other conditions.

Data from the bear studies have led the scientists to devise a very low-protein, low-fluid diet for patients with kidney failure. Studies on patients without kidneys at the Mayo Clinic have shown that such patients can go ten days instead of the usual three before needing another treatment with the artificial kidney machine.

The study team includes physicians, veterinarians, medical students, and technicians. In working with the bears, they brave not only Minnesota's harsh winters but also the dangers involved in arousing and then sticking needles into the 500-pound animals to inject radioactive substances and to withdraw blood samples at scheduled intervals thereafter.

By testing the blood samples in a laboratory to follow the radioactive decay pattern, the scientists learn how the bear can use the stored foodstuffs—fats, proteins, and sugars—so precisely that they do not need extra calories during the period of "hibernation" from December to March.

Although wild bears sleep up to five months in the winter, they are not true hibernators like woodchucks, ground squirrels, and many reptiles. The bear's temperature drops by just 4 degrees, whereas those of true hibernators plummet more than 60 degrees. Unlike many small hibernators that sleep so soundly they can be picked up and tossed at will, a bear in winter sleep will arise at the slightest noise, charge at a visitor, and, if sufficiently concerned about the disturbance, even move its den.

When Mrs. Dianne Wellik, a research assistant, and a visitor approached the cage with a syringe and needle attached to a long pole, the bear stood on its hindlegs, grabbed the mesh with its front paws, and spat. Bears can be subdued for purposes of the experiments only when quieted with injections of a combination of muscle-relaxing and anesthetic drugs.

When the drugs took effect, the team carried the bear from the cage out into the open where it was placed on its back.

Mrs. Wellik ran her fingers over the bear's lower abdomen until she felt a pulse that guided her to the femoral vein. Then she took a large syringe and needle and withdrew samples of blood. A few minutes after the tests were completed, the bear was carried back to the cage, and placed in a sitting position to help prevent the animal from developing pneumonia.

The bear experiments had their origin in a passing remark made to Dr. Nelson several years earlier by another doctor who had just returned from a medical meeting where he heard about the bear's hibernating characteristics. Dr. Nelson, a physician who also has a Ph.D. degree in physiology and who specializes in nutrition, was puzzled: how could hibernating bears survive while accumulating toxic waste products?

Dr. Nelson had no plans to study bears until he received a surprise call from his technician, James Penner. Three bears were available, said Mr. Penner. Having overheard Dr. Nelson's comments, Penner had taken it upon himself to seek assistance from Dr. Paul E. Zollman, a veterinarian who cares for the animals used in experiments on the farm.

Dr. Nelson, who enjoys telling stories about his bears as much as doing the research, laughed as he recalled how the first bear arrived at the Mayo Clinic from the Upper Peninsula in Michigan on the back seat of a Volkswagen. The bear had been calmed for the 300–mile trip with injections of a muscle relaxant.

"Everyone who works with bears gains an immediate respect and love for these animals," Dr. Nelson said. "They go without food and water, yet the females can incubate fetuses and then nurse cubs during winter sleep. Other animals can't find the bear because by producing no urine or feces he leaves no odor."

When the team first began testing the bears, Dr. Nelson said that "Dr. Zollman would put a syringe containing an anesthetic on the end of a stick, walk in, back the bear into a corner of the cage, stick him, and rush out, escaping the animal's charge. We couldn't ask Paul to risk his life all the time, so we used our ingenuity, got a longer pole, and learned how to inject the bear while standing outside the cage."

Some bears fail to hibernate in the wild, and the Mayo team was told that the bears would have even more difficulty hibernating in captivity. The researchers learned that the bear would hibernate only if they denned it in a quiet, isolated spot. The Mayo bears den on the farm in what was a root cellar.

There, Dr. Zollman has matted straw in flattened, oval, corrugated metal culverts of the type used in road construction, and erected a set of protective bars.

When the studies are done in the hibernating season, Mrs. Wellik keeps warm by carefully huddling against the anesthetized bear's fur and by wearing battery-powered socks. But her hands are numbed by the cold because she cannot use gloves when she guides the syringe and needle into the bear's veins to inject the radioactive materials and to withdraw the samples that may need to be taken every few minutes.

The researchers work strictly with private funds and donate personal lecture fees to further their bear research. Because the logistics make it too difficult to work with more than three bears at any one time, bear research has proceeded more slowly than it might have if other animals were used.

"People ask why we study such a large, dangerous, mean animal and not the smaller hibernating woodchuck," Dr. Nelson said. "But there are advantages working with large animals. With a small animal it is difficult to do tests on small amounts of blood and tissues. Some studies we do on the bear would be impossible to do on a smaller animal because we would bleed it to death taking the samples we need. The bear is so large your chance of isolating substances is much better."

The Mayo team usually does an experiment one winter,

repeats it during the nonhibernating season for comparative purposes, and then confirms results the following winter. Each year a new experiment is added. The team plans well ahead because a missed opportunity means waiting another year.

Interpretation of the data collected reflects hours of thought and observation. Dr. Nelson said he was taught "that doing research is like receiving a cablegram. From just a few words you have to figure out what's going on back there." By such insights, Dr. Nelson said, "we know what happens, how it happens, but we don't know what switches on the gears."

Over the years, Dr. Nelson's team has learned that the bear's reaction to hibernation resembles the human's in starvation "except that the bear does it perfectly." Starving or bedridden humans burn up stores of fat but they also lose protein as muscles waste.

The bear, however, tolerates hibernation very well. For about one month before going into winter sleep, the wild bear becomes ravenous, eating for twenty out of every twenty-four hours, increasing his daily caloric intake up to 20,000 from about 7,000, and gaining more than 100 pounds.

Then, once the energy input stops and sleep begins, the bear's biochemical reactions become delicately balanced. The fasting bear makes just enough water from its fat stores to stay hydrated. Normally the bear, like other animals, manufactures and breaks down proteins at a constant rate in a process called "protein turnover." However, the researchers found that protein turnover speeds up five times during hibernation without increasing the amount of protein in the body at any one time. This conservation program provides the bear with as much protein at the end as at the beginning of winter sleep.

The process is so efficient that the bear does not form

excess amounts of urea, the waste product of protein breakdown, which is excreted in urine and which becomes toxic when large quantities build up as a result of damaged kidneys. During hibernation, the bear's two kidneys produce just a trickle of urine and that is reabsorbed into the blood through the bladder wall.

The bear's adjustment to hibernation is so sophisticated that results of most standard laboratory tests of blood samples from the nonhibernating and hibernating periods are the same. The only exception is that cholesterol and other fats rise during hibernation.

When the bear wakes up in the spring after a five-month fast, he still is not hungry. "We can't get him to eat for two weeks even when we serve him food," Dr. Nelson said.

The researchers have found that bears that cannot hibernate starve to death just like a human or other animal. And because the researchers have tried unsuccessfully to get bears to hibernate in the summer by putting them in a dark, cold room mimicking winter conditions, they suspect the phenomenon is controlled by a hormone that the brain's hypothalamus makes each fall and winter.

Dr. Nelson speculates that discovery of a hormone controlling winter sleep might help patients with kidney disease as insulin helps people with diabetes. He also theorizes that because the bears lose their appetite in the face of food, such a hormone might prove helpful in treating obese people. At the same time, it might help prevent malnutrition among the starving peoples of the world. "The bear goes through both extremes—starvation and obesity—each year, but the bear treats obesity as no human can," Dr. Nelson said.

18

UNRAVELING THE SHAPE OF AN ENZYME

Malcolm W. Browne

CAMBRIDGE, Massachusetts—Some revolutionary tools for probing life's secrets are taking shape under the hand of William Nunn Lipscomb Jr., winner of the 1976 Nobel Prize in chemistry.

Dr. Lipscomb's new work, in the view of many colleagues worldwide, is so significant that it could well win him a second Nobel Prize. A tall, affable Kentuckian, he smiles self-deprecatingly at mention of Nobel Prizes, but acknowledges that his current research is the most important he has done.

The three decades of work that led to Dr. Lipscomb's Nobel Prize were mainly with a class of chemical compounds called boranes, molecules made of boron and hydrogen. His current work is chiefly with a regulatory enzyme called aspartate transcarbamylase, or ATCase, one of the key triggers of cell division and growth in all living things.

Biochemists, biologists, and doctors are watching with great interest, since the work of the Gibbs Laboratory of Harvard University, over which Dr. Lipscomb presides, promises to shed light on such vitally practical matters as cancer cell division.

Among scientists who have expressed special praise for Dr. Lipscomb's current work is one of his former students, Dr. Raoul Hoffman, a Cornell University chemist. "His work is particularly remarkable now, not only because of its intrinsic value but because it is so unlike his earlier work in boranes," Dr. Hoffman said. "It is, in my view, as likely to win him a Nobel Prize as was the borane work."

Paradoxically, Dr. Lipscomb is a physical chemist, not a biologist or even an organic chemist. The ideas with which he works have to do with geometric shapes and transformations, quantum mechanics, and relativity theory, seemingly more in the realm of lifeless abstractions than that of microbes or living tissue.

Dr. Lipscomb, fifty-seven years old, has contended throughout most of his career that biology cannot be fully understood without explanations from mathematics and physics.

"The old boundaries between the different branches of science are breaking down fast," he said in an interview. "For many years, the biochemists have been at the fore with their dramatic discoveries in DNA research and so forth, but the major breakthroughs from here on, I believe, are likely to come from the fundamental branches, physics and physical chemistry."

Dr. Lipscomb had been impressed by the knowledge gained from earlier studies that the way atoms interact within a molecule depends not only on the classical ideas of electron bonding, involving the interchange of electrons, but also on the respective positions of the various atoms in three-dimensional space.

As the physical shapes of certain molecules are warped or distorted, it was found, the electronic bonds between their constituent atoms may be radically changed, altering the chemical behavior of the whole molecule.

Such phenomena could explain some of the mysteries of the behavior of complex molecules in living systems, Dr. Lipscomb believed, and he decided a logical start could be made with ATCase. Like the hundreds of other enzymes, the substance catalyzes chemical reactions without taking part in them. The physical shape of the molecule could explain how it triggered cell division, he hypothesized.

The task of devising a three-dimensional map of the ATCase molecule now occupies most of Dr. Lipscomb's research time. He also teaches full-time for four months a year and travels to participate in many symposia in other months.

Superficially, there seems to be an enormous gulf between the lifeless, exotic boranes and such a huge, complicated molecule as ATCase, which is essential to life.

But for Dr. Lipscomb, enzyme research is a logical outgrowth of his borane work, which is also continuing, because the computer techniques he developed to determine the latter's structure can be extended to investigations of the former, which is far more complex.

"I periodically decide to give up borane research, imagining that there is little more to discover," he said. "But I have never abandoned boranes for more than six months, because some new idea always brings me back, making me realize how much more there is to be learned about chemistry in general from boranes than I had previously imagined."

For many years, fellow chemists regarded Dr. Lipscomb as mildly eccentric, not merely because he affects bright checked shirts and string ties, or because he has a passion for chamber music and great skill with the clarinet.

For an aspiring young chemist in the 1940s, few subjects could seem so unlikely to result in major scientific discovery (or practical applications) as boron chemistry. Boron research at the time seemed to most chemists an utter waste of time. But Dr. Lipscomb had become intrigued by the subject and stimulated by the thinking of his teacher and mentor, the Nobel laureate, Linus Pauling.

"The more I thought about boranes," Dr. Lipscomb said,

"the more I came to realize the importance of molecular relationships between structure and function."

Thus, Dr. Lipscomb explained, the same atoms arranged in a three-dimensional molecule may behave quite differently when twisted or pushed out of shape with respect to each other. Such proved to be the case with the boranes, strange molecules shaped like cages, some having twenty faces or more.

Dr. Lipscomb, and the dozen or so doctoral and postdoctoral students who perform the actual experimental work for him, determined the shape of many of the boranes using a device called the x-ray diffractometer, a table-top x-ray machine for looking at tiny crystals.

But before many of the new boranes were actually discovered, Dr. Lipscomb had predicted their existence with the aid of higher mathematics and a children's construction toy called "D-Stix," from which complicated models took shape.

Such tools had been refined to a high degree in his borane work when Dr. Lipscomb decided to take on ATCase, a winding, convoluted tangle of amino acids and related substances. It had long been established that the enzyme was related to cell division, but an explanation of how the enzyme's action is triggered has eluded biochemists.

Dr. Lipscomb reasoned that since chemical function always depends on structure, an exact knowledge of the shape of the ATCase molecule would ultimately answer the question. A key part of that question is how a chemical signal impinging on one of the extremities of the molecule is transmitted over the relatively huge distance of 40 Angstrom units (40 ten-billionths of a meter) to its "active center."

The "active center," the part of the molecule that causes it to trigger division in other cells, is in a cavity indented in the center of its structure.

To learn about the matter requires determining the spatial

relationship of every atom in the molecule to every other atom in the vast structure. The magnitude of the task, which is likely to continue for several more years, is evident from the molecular weight of ATCase: more than 300,000. (The molecular weight of the simplest molecule, hydrogen, is about 2.)

It is slow work, but improvements in the tools have helped. The circular patterns of dots that result from Dr. Lipscomb's x-ray diffraction photographs are now scanned automatically and the results translated into computer language.

Computers have been taught to draw topographical maps using the data, maps that look like those made by geologists to depict mountains and valleys. The contours derived from the x-ray data, however, mark the positions and spatial relationships of atoms and clusters of atoms inside a molecule.

The resolution, or focus, of the pictures thus obtained has been described by Dr. Lipscomb's admirers as a triumph of scientific technique.

Similar techniques enhanced by Dr. Lipscomb and his coworkers have resulted in a wholly new way of looking at chemical bonding, with the realization that a single electron can bind as many as three atoms together within a molecule, not just two.

"To give you an idea of how radically chemistry has changed over the years," Dr. Lipscomb said, "I can say that I spend practically no time in a laboratory. At the Gibbs Laboratory we do no chemical analysis and practically no synthesis.

"But by contrast, I probably use more time on Harvard's computers and other computers available to us than any individual—huge hunks of computer time.

"It's a far cry from my own antecedents. I became interested in chemistry when my mother gave me an A. C. Gilbert chemistry set. But today the real research is in ideas—even intuition—expressed, usually, in the special languages with which mathematicians and computers communicate."

Dr. Lipscomb communicates many of his ideas to colleagues and students with visual aids, models, and three-dimensional photographs and drawings that must be viewed with stereoscopic devices.

One of his key assistants is a former Radcliffe College tennis coach, Jean Evans, an artist who works at the Gibbs Laboratory drawing the diagrams and sketches needed to make the chemist's ideas easily intelligible. Dr. Lipscomb, on the other hand, has become an ardent tennis player.

Held in great affection by colleagues, who invariably address him as "Colonel" in recognition of his Kentucky origins and honorary colonel's rank, Dr. Lipscomb teaches that the creative process should be kept apart from purely practical endeavor.

"For me," he said, "the creative process, first of all, requires a good nine hours' sleep a night. Second, it must not be pushed too hard by the need to produce practical applications. Without fundamental basic research, we quickly run out of the discoveries that give the engineers and therapists and other practical people the material to advance.

"In some ways, our crash approach to finding a cure for cancer has been mistaken, I believe," he said.

"The funding should provide more emphasis on basic research. As for cancer itself, I don't believe a single cure will ever be found, because it is not one disease but hundreds of different ones all described by the same name.

"But through basic research we can chip away at each disease individually. And, which for me is equally important, we can increase man's fundamental knowledge of the universe of which he is part."

19

A SEARCH FOR A NEW MALE CONTRACEPTIVE

Jane E. Brody

NEW YORK—Scientists have tried for decades to develop a "modern" contraceptive for men. A drug was found that stopped sperm production, but it turned men's eyes red and caused them to vomit when they drank alcohol. The female hormone progestin also was shown to halt the manufacture of sperm, but it suppresses libido, necessitating a daily injection of male hormone to restore the sex drive.

Testicles have been bombarded with ultrasonic waves, infrared rays, and laser beams. But in each case the resulting suppression of sperm was unpredictable, varying from man to man, even day to day.

And so, for David M. Phillips and hundreds of other scientists interested in male reproductive biology, it's back to the laboratory if there is ever going to be a revolution in male birth control comparable to what the pill did for women.

"We're not at a dead end," the thirty-nine-year-old cell biologist remarked, gesturing toward a wall papered with greatly magnified pictures of sperm at various stages of development. "There are plenty of things to study. But we need to do a lot of work at the basic level before we can come up with some rational and effective new techniques. Otherwise, we will have nothing but shotgun approaches riddled with potentially serious hazards."

Dr. Phillips explained that developing twentieth-century

contraceptives for women was comparatively easy. There are a lot of stages at which female fertility can be shut off—by interfering with the monthly release of an egg, blocking transport of the sperm to the egg, stopping implantation of the fertilized egg in the womb. These led to development of oral contraceptives, sperm-killing foams, and intrauterine devices.

"Fifty years ago when it was first discovered that female hormones operated on a monthly cycle, it occurred to scientists that it might be possible to develop a hormonal contraceptive to interrupt the cycle," Dr. Phillips said.

"But there is no comparable cycle in men. Women release one egg a month; men make an average of 300 million sperm each day. It's a much bigger target, and you have to stop every one of them to have an effective contraceptive."

He added that the biology of male fertility was also different. The hormone, testosterone, that stimulates the production of sperm in the testes, also controls libido.

"If you knock out testosterone, you have the effect of castration, and this is not an acceptable contraceptive," Dr. Phillips remarked. In women, however, ovulation can be stopped by maintaining normal high levels of progestin, but it is estrogen that influences female libido.

A number of nonhormonal chemicals have been discovered incidentally that stop sperm production, but they are risky because they damage genes and may cause cancer and birth defects, among other toxic effects.

Therefore, as Dr. Phillips and others see it, the real hope for a safe and effective male birth control pill lies in deciphering the specific events in the production of working sperm to identify vulnerable links in the chain and develop safe ways to interrupt them.

Dr. Phillips is a microscopist at The Population Council, a nonprofit research institute. The council's Center for Biomedical Research, where Dr. Phillips has his lab, is located at Rockefeller University in New York. The scientist spends his days studying sperm, fast-swimming cells that are specially designed for their one task in life—to deliver the male's genes to the female's egg. Sperm spend about three months "growing up" to expend all their vital force in but a few hours.

Dr. Phillips wants to know the precise details of their origin in the germ cells of the testes; their maturation in the 800 or so coiled seminiferous tubules in the testes and the epididymis next to each testis, and the mysterious but crucial change they undergo in the female, called capacitation, that finally allows them to fertilize an egg.

"If one could safely interfere with any one of these steps, we would have an effective contraceptive," Dr. Phillips pointed out.

Specifically, Dr. Phillips studies the subtle structural changes accompanying the various stages of sperm development. To determine which of these changes is critical to the viability of the ultimate fertile sperm, he examines sperm of many species, among them rats, mice, butterflies, beetles, opossums, and occasionally men. These comparative studies reveal which developmental features are common to sperm generally and therefore probably represent essential, rather than incidental, characteristics.

Man is only a sometime subject of his studies because, Dr. Phillips explained, except on the rare occasions when he obtains testes removed during surgery, "we can only get human sperm from ejaculates, after they have already matured. We can't go around castrating men so that we can study the maturation of their sperm."

Dr. Phillips said that as a scientist he "got hooked" on sperm in the first place because he was struck by the physical beauty of some fly sperm he saw under an electron microscope during his graduate studies in 1962.

He continues to be impressed by the esthetic qualities of his study subjects. All sperm have basically the same job to do, and within a species they are remarkably consistently alike. But unlike other types of cells, they vary in intriguing ways from species to species.

As Dr. Phillips described it, "The Chinese hamster has about the biggest sperm of any animal studied. Rats have tremendous sperm, much bigger than the mouse or man. Elephant sperm are smaller still. The head of a rat sperm is sickle-shaped; squirrel sperm are shaped like a spoon; the platypus sperm looks like a pencil; the bull, bat, and rabbit have sperm with flattened heads that resemble lollipops, and the human sperm looks like a fat lollipop."

The reasons for these different shapes and sizes are unknown, but they serve to make Dr. Phillips' work interesting.

"I like to paint and I'm somewhat of a frustrated photographer, but when I sit at my microscope and look at the beautiful arrangements and shapes of these cells, I don't need to paint." However, he has taken thousands upon thousands of photographs of sperm, some of which resemble patterns on wallpaper or Finnish fabric.

Dr. Phillips began his in-depth studies of sperm at Harvard Medical School as a postdoctoral researcher under Dr. Don W. Fawcett, who in the last twenty years has probably done more to elucidate the structure of sperm than anyone since Antony van Leeuwenhoek first described their microscopic appearance to the Royal Society of London 301 years ago.

The sperm is a streamlined cell, stripped down to serve as an efficient transport system. It has a head and a tail. The head contains a tightly packed nucleus of genetic material—the "payload," as Dr. Phillips calls it—and very little cytoplasm.

The mature sperm cell has no machinery for making protein, but it does have a series of "powerhouses"—organelles called mitochondria—that convert sugars in the environment into the energy the cell needs to swim. Its "motor" is a whiplike tail called a flagellum.

The sperm head is capped by a "warhead" called an acrosome containing enzymes that can dissolve the layers of cells and mucus that surround the egg, enabling the sperm to deliver its payload of genes into the egg.

Although much is known about the gross structure of mammalian sperm, Dr. Phillips is trying to determine what small structural changes take place in the course of sperm development that might prove to be biochemical Achilles heels. He studies sperm at various developmental stages by removing them from different parts of the animals' reproductive system and imbedding the cells in plastic. They are then sliced by a machine that cuts sections 600 angstroms thick, allowing a single cell to be sliced into 100 pieces.

Using an eyebrow hair at the end of a toothpick to separate the slices, Dr. Phillips stains them with heavy metals and examines them under his $100,000 transmission electron microscope.

He also studies the surface changes in sperm cells under a scanning electron microscope, which permits three-dimensional viewing of intact organisms, in this case the whole sperm cell. And he makes platinum and carbon replicas of the cell's surface—a kind of metallic bas relief—which he can examine under the finer magnification of the transmission electron microscope.

But Dr. Phillips leaves the biochemical studies to others. "I'm trying to figure out what happens structurally during maturation and capacitation," he explained. "When I find something, I publish my results and maybe someone else will be working on a related aspect that fits together with my findings.

"That's how science is done—a lot of people adding little pieces together until you have the whole puzzle. There are no more Renaissance-type scientists who do everything themselves. I find something, someone else finds something based on what I found, and so forth and so on, until one guy develops the contraceptive—and he gets the Nobel Prize."

As Dr. Phillips outlined them, the possible approaches to intercepting the sperm's mission include the following:

Immunological methods, such as developing antibodies to a protein on the surface of the sperm.

Interfering with the release of enzymes from the acrosome or with capacitation in the female.

Causing the sperm head to separate from its tail. A recently deceased English bull ejaculated only tailless sperm, suggesting the feasibility of this approach.

Preventing the breakup of the bridges of cytoplasm that connect sperm in long strings as they are forming. A necklace of sperm could hardly find its way to and into an egg.

"But none of these can be done," Dr. Phillips cautioned, "without a lot more basic work. It's easy to speculate about the possibilities, but if you don't have a firm basic foundation for what you're doing, you end up with a lot of poor clinical research that could be hazardous to people.

"In this field of contraceptive development, it's easy to find desperate people who are willing to be human guinea pigs. We must guard against being pressured into developing and testing something in man before we fully understand what we're doing."

V
Exploration of Life—II

ANIMAL BEHAVIOR, ECOLOGY, LIMNOLOGY

Dr. F. Herbert Bormann counting blossoms on a young pin cherry tree in Hubbard Brook Experimental Forest. This is done to determine reproductive powers.

Dr. William B. Jackson, left, studying dead rats in camp at Enjebi.

20
A FOREST AS A LIVING LAB
Bayard Webster

WEST THORNTON, New Hampshire—From the center of this small country village, as from the bottom of a fluted bowl, the green hills of the White Mountain National Forest rise abruptly to the sky wherever the eye turns. From the rim, the richly forested land seems to tumble downward in waves and ridges, forming nearly a score of valleys and small watersheds whose streams feed into nearby Hubbard Brook.

This is where, eons ago, the glaciers gouged and rounded off the earth's surface to form the graceful contours, now carpeted with green, that are part of one of the great forests of North America.

But in this particular section, known as the Hubbard Brook Experimental Forest, a sharp-eyed visitor may notice some strange aberrations among the verdant hills: a scalped mountaintop here, a pie-shaped, faded brown clearing there or, farther along, a tract resembling a giant yellow-and-green ladder.

These scars and perturbations are not the result of commercial loggers, or of insect depredations, or of disease. Paradoxically, they are the work of human researchers who are deliberately manipulating the forest in an effort to find out how man and the woods that his ancestors lived in years ago may coexist more fruitfully and harmoniously in the future.

Their collective work in the White Mountain National Forest is believed to comprise the most comprehensive and significant study of forest watershed ecosystems that has ever been undertaken.

One of these scientists is Dr. F. Herbert Bormann, a stocky, energetic, Manhattan-born ecologist who is former president of the Ecological Society of America. He is currently Oastler Professor of Forest Ecology at the Yale School of Forestry and Environmental Studies in New Haven.

The fifty-three-year-old Dr. Bormann, whose black hair is accented by snow-white sideburns, is one of a relatively new breed of ecologists—those who follow a systems approach in their scientific discipline.

But applying systems research to a forest watershed (a sloped, woodland area drained by a stream) is no easy task, Dr. Bormann said. "It's not like studying how two things react in the lab. There are so many variable factors like air currents, storms, rain, sunshine, and animal activity that you can't control."

In the past, ecology was often considered to be simply the study of a single species in relation to its immediate environment. Now, the predisposition of a growing number of ecologists is to examine whole ecosystems—forests, prairies, marshes, rivers, and the like—and how all living species, including man, fit and work together in their particular ecosystem.

Contrary to the image, held by some, of ecologists as little more than rabid preservationists, political obstructionists, or spokesmen for environmental pressure groups, the majority of ecologists are serious teachers and researchers in a discipline whose scope and contours are still evolving.

They work in the laboratory and, increasingly, in the field where they may prove their lab findings or, conversely, where they may make discoveries to be confirmed later in the lab.

On a recent weekend, Dr. Bormann led a small group into

one of the watersheds in the 7,500-acre Hubbard Brook Forest. It was his first visit of the year to his outdoor laboratory, some 250 miles from his office and classrooms in New Haven.

He first came to these woods more than a dozen years ago in the company of a colleague and friend, Dr. Gene E. Likens, a professor of ecology at Cornell. They shared a mutual curiosity about the sylvan ecosystem.

How, they wondered, does the forest cycle its nutrients, the chemical and mineral elements necessary for its life? What does the forest take from the air, the rain, and the animals it harbors? What is the chemistry of its streams?

If they could find the answers to these questions, they reasoned, they might learn how better to manage and preserve some of the forests that cover 38 percent of the earth's land surface and produce about 20 percent of its oxygen.

They obtained permission from the U.S. Forest Service to turn one of the watersheds into a living laboratory, and later received grants from the National Science Foundation to help fund their work.

Now, thirteen years after the initial request, nine watersheds are being studied here by scores of scientists from universities, industry, and the Forest Service. Their research involves clear-cutting (a controversial logging method by which every single tree in the cutting area is felled), strip-cutting (clear-cutting in strips, causing a ladderlike effect), and applying herbicides to restrict forest growth.

They also measure tree bulk, count bird populations, assay stream water chemistry, examine soil samples, analyze the forest air, and study the way a cut forest regenerates itself.

As members of the group—Dr. Bormann, fellow forest ecologists from Cornell, two Yale forestry students, Dr. Bormann's daughter, Becky, newly graduated from Duke, and a reporter—put on their hiking boots, one of them was reminded of the time-tested method of identifying at a distance the different types of scientists that one runs across in the forest.

Bird people (ornithologists) are always looking up, he remembered. Plant people (forest ecologists and botanists) are always looking down, and stream people (hydrologists and limnologists) have wet boots.

With the sun shining brightly through the canopy of maple, beech, and birch, Dr. Bormann led the climb into Watershed 2, an area that had been completely clear-cut several years ago. At the foot of the watershed, a stream bounced off the rocks before being slowed up by a weir that had been installed in the 1950s by Forest Service hydrologists.

By constructing a weir—a watertight, damlike structure that acts as a settling basin and directs stream flow over a V-notch measuring section—at the bottom of each watershed, Dr. Bormann explained, it became possible to sample quantitatively the stream flow. This was done for several years to establish baseline, or reference, data on the amount of minerals and chemicals—such as calcium, iron, magnesium, nitrogen, phosphorus, potassium, and sodium—found in the water in the stream at the bottom of the watershed.

After the baseline data had been established, the first major experiment was conducted in 1966—the cutting of every single plant and tree on Watershed 2, a thirty-nine-acre plot. Following the clear-cutting, a herbicide, Bromacil, was applied for a three-year period to stop all plant regrowth and to prevent any new vegetation from growing.

As a result of the denuding of the forest, researchers found that the quality and quantity of the water that issued from the soil into the stream had changed dramatically. Their laboratory analyses showed that the volume of particulate matter had increased significantly and that the losses of nutrients that would ordinarily have been absorbed by tree roots, leaves, and branches ranged from three to sixteen times greater than in stream water in an undisturbed forest.

In addition, the amounts of nitrates in the stream rose from 2 milligrams a liter to 90 milligrams a liter, an amount far exceeding U.S. Public Health Service limits. And the volume of stream water at the bottom of the mountain increased by 40 percent in the first year following the clear-cut, falling to a 30 percent increase in the following two years.

In 1969 the herbicide application was stopped and the denuded forest was "released"—allowed to regenerate. Now, six years later, it displayed a vigorous young growth of beech, maple, and birch, and the stream-water chemistry had approached its precutting levels.

Assessment of the project is still going on, but researchers point out that the evidence indicates that clear-cutting could provide increased water flow to water-short cities downstream from the forest.

Increased water flow, however, would come at the expense of lower nutrient levels in the land and, therefore, less vegetative productivity, so there might be no net advantage to man. "This is an example of the complexity of these systems," Dr. Bormann said. "It shows you have to pay a toll when you do something to the land."

Pushing their way through thick stands of raspberry bushes and pin cherry trees, the group moved on to Watershed 4, where more recent research was being conducted.

Here the slope had been divided into eighteen 25-meter-wide strips laid out at right angles to the fall line of the mountain. Beginning in 1970, every third strip, for a total of six, was clear-cut. Two years later another six strips were cut in the fifty-year-old forest. And in 1976 the last standing strips were cut.

As they climbed the mountain, Dr. Bormann and his

group outlined with bright yellow string and plastic flags a series of small plots in the strips that had started to regenerate plant growth after they had been cut in 1970. Later the students would return to harvest the plots and take the plant material back to Yale. There they would analyze the vegetation under the supervision of Dr. Bormann.

The plants are first dried in ovens to remove their moisture. After dry weighing they are chopped up and their chemical and mineral nutrients are removed by liquid solvents. The extracted nutrients are then heated in a flame and analyzed by a process known as atomic absorption spectrophotometry, in which identification of each element and its concentration are determined optically through the use of photoelectric cells.

Using these data, Dr. Bormann and his students determine how quickly or slowly the plants (young trees) are taking up nutrients from the soil and, accordingly, the rate of forest growth in the strip-cut area. This rate is then compared to that of other plants that grew in different areas and were at different stages of their development.

In the strips that had been denuded four and one-half years ago, the regenerative powers of the forest were obvious. Dense, 10-foot-high stands of fast-growing pin cherry trees had already been produced and slower-growing species such as yellow birch, sugar maple, and beech were flourishing.

Dr. Bormann noted that the strip-cut method of harvesting timber seem to encourage the growth of yellow birch, a highly marketable wood. This might prove to be a boon to commercial foresters in addition to preventing a massive runoff of nutrients and disturbances of the forest ecosystem's natural functions that occur after total clear-cuts.

"The more we study the forest," he said, "the more we learn about why we should be gentle with it. Not many people know what a forest is, what it does, and the ways in which it contributes to mankind, in addition to producing wood."

Scientific studies, including several that he and Dr. Likens conducted, have shown that mature forest ecosystems act to minimize flooding and soil erosion. They also reduce temperature extremes, yield chemically clean water to streams and ground water reserves, and act to filter the air of such gaseous pollutants as carbon monoxide, sulphur dioxide, and ozone.

21

WADING IN THE NAME OF SCIENCE

Bayard Webster

WEST MARLBORO TOWNSHIP, Pennsylvania—Propping herself against the trunk of her car, the stocky, dark-haired woman pulled on a pair of well-worn hip boots, grabbed a long-handled dip net and walked briskly toward a nearby creek, a clutch of fourteen graduate students, all similarly booted, in her wake.

They were setting out to examine the aquatic life in the stream, a carefully monitored living laboratory that meanders through the Stroud Water Research Center here, believed to be the world's most comprehensive facility for the study of fresh-water streams and ponds.

The woman at the head of the group was sixty-eight-year-old Ruth Patrick, one of the country's leading limnologists —those scientists who spend a good part of their lives studying lakes and streams. Dr. Patrick's latest excursion into the clear Chester County stream was one of the several thousands of times she has pulled on her boots and gone wading in the name of science. "I don't believe in armchair investigating," she said as a chill wind riffled the surface of the creek.

The results of her expeditions here and in many parts of the world have brought her renown as one of the world's experts on the microscopic plants known as diatoms. These one-celled organisms, whose species number in the thousands, make up the majority of the algae family. They are found in both fresh and salt water and in moist places in every sector of the earth.

Unlike other algal forms, they have hard, shiny outer shells of silica and, by the process of photosynthesis, produce a kind of oil instead of the usual cellulose products—starches, sugars, and carbohydrates—that most plants manufacture. Dr. Patrick has learned to use diatoms (pronounced dýe-a-toms) as water pollution indicators. She devised a system that relates the types and numbers of diatoms present in a stream or lake to the type and extent of pollution, enabling her quickly to identify contamination problems. This method is now used in many parts of the world to help determine water quality.

Although she was not going to make her students go through the methodology of her pollution identification system, she was taking them on one of their first field trips, gathering samples of animal and plant life, including diatoms, and then examining and classifying them under the microscopes in the laboratory. The scientist-teacher was on the banks of her home stream now, that belongs to the limnology department, which she founded, of the Academy of Natural Sciences of Philadelphia, of which she is chairman. She was looking for new discoveries and at the same time hoping to transfer some of her knowledge to her students—future biologists and limnologists from the University of Pennsylvania, where she is professor of biology.

As she led the way, she and her students scooped up samples of insects, algae, moss, and mud from the stream bed and placed them on screens of fine-mesh sieves. Several of the group watched through a hand lens as a minuscule black fly emerged from its pupal casing and unsteadily began to walk.

Pouring samples of stream water into small plastic bags, the students walked back to the Stroud laboratory building, took their boots off, and prepared slides of the samples for microscopic examination. Moving from scope to scope, Dr. Patrick quickly identified the organisms when a student was stumped, reacting spiritedly ("Oh, there's a cladophora!") when a species of her favorites, the diatoms, appeared. When magnified 440 times, some kinds of diatoms resembled tiny boats that occasionally bumped into each other.

Diatoms range in length from one micron (one millionth of a meter) to a millimeter in length and are constructed somewhat like hinged pillboxes. Their shapes vary from thin rods to oval and pancakelike forms. The thousands of separate species are identified by differences in size, shape, and the infinitely tiny radial lines, ridges, spines, hairs, and nodules on the organisms.

Dr. Patrick became interested in diatoms at a tender age. Her father, Frank Patrick, a lawyer by trade and a biologist by avocation, was an admirer of the tiny diatoms, often called the "gems of the sea." As a treat for his daughter he would lift her onto his lap and let her peer through his microscope at the jewellike forms. "I was hooked on diatoms at the age of seven," Dr. Patrick remarked.

Pursuing her interest into maturity, she got her masters and Ph.D degrees in biology at the University of Virginia in Charlottesville. She began her affiliation with the Academy of Natural Sciences (which had the best collection of diatoms in America) in 1933 and in 1947 established and headed its limnology department.

Historically, diatom remains have been scientifically useful in charting the advance and retreat of glaciers and in tracing the prehistoric conditions of ancient lakes. In some sections of the world, huge deposits of the dead plants constitute diatomaceous earth and are mined commercially for use in thermal insulation and ultrafine filters, as filters for paints and plastics, and as a polishing agent for lenses and mirrors.

Live diatoms are at the bottom of the food web, where they provide a vital part of the diet of tiny insects, mollusks, and fish, contributing essential vitamins and other nutrients. But in the eyes of Dr. Patrick and other limnologists, living diatoms now have another important use for mankind, for they play a key role in helping to combat the earth's most serious pollution problem—the contamination of its oceans, lakes, and streams.

A few decades ago it was generally known among aquatic scientists that a few kinds of diatoms seemed to favor certain types of water environment. With her extensive knowledge of the different types of diatoms, Dr. Patrick set out in 1948 to correlate these relationships more precisely.

In the company of many scientists, including a chemist, a bacteriologist, and animal and plant experts, Dr. Patrick examined a Pennsylvania stream, Conestoga Creek, for basic data on the diatom-environment relationship. The creek, a small river that winds through Lancaster Country, was considered classic for their purpose—it suffered from almost all known types of pollution: sewage, fertilizer and pesticide runoff, metals from heavy industry, and toxic substances.

The scientists mapped the various sections of the river and its stream bed according to the various types of pollution, and then identified the types of plant and animal communities found in each of those sections. Dr. Patrick found that some diatom species thrived in water that was heavily polluted with organic contaminants such as human sewage, animal manure, and crop fertilizers. Others flourished in pollutants of a different nature, such as mineral or chemical.

By linking up the specific kinds of diatoms with the specific kind of water environment they thrived in or were absent from, it became possible to determine the type of pollution simply by examining and classifying the diatoms found in a stream. Further refining this system, Dr. Patrick learned to examine a sample of stream water under the microscope and determine from the types and numbers of diatoms present not only what kind of pollution was present but how severe it was.

To make the collection of diatoms easier, Dr. Patrick invented the diatometer, a plastic box containing microscope slides that, strategically placed in a stream, capture the maximum number of the organisms. These devices are now in more than seventy sites in the United States as monitors of water pollution.

Dr. Patrick's research on diatoms and streams has taken her to bodies of water in Canada, South America, Ireland, Mexico, Thailand, Austria, and almost every part of the United States. Recently she has worked at the Stroud Laboratory on the role of trace elements in the nourishment of diatoms. For her work in these fields and her published studies of diatom taxonomy (classification), Dr. Patrick last year was awarded the $150,000 John and Alice Tyler Ecology Award—the world's richest prize for scientific achievement.

Dr. Patrick, who uses her maiden name professionally, lives in a tree-shaded home in Chestnut Hill with her husband, Dr. Charles Hodge, a retired entomologist, and her son, Charles, a medical student at Temple University. She is eager to give up her job as chairman of the board of the Academy of Natural Sciences. "I want to get back to research full-time," she said.

Dr. Patrick, in cooperation with Rutgers University, is currently studying the effects of petroleum in estuaries of the Delaware River. Financial support from industry for part of her research work, plus the fact that she is on the boards of DuPont and the Pennsylvania Power and Light Company, have elevated the eyebrows of some scientists who feel that scientific water pollution research should not be funded by industries that themselves are polluters.

But Dr. Patrick, who feels that she had to work harder and longer than her male peers to reach her present eminence, believes in "being involved with the people who are causing the problems." In an interview she remarked, "Industry has got to face up to the truth—they've got to be willing to admit it when they do something wrong. We have to develop an atmosphere where the industrialist trusts the scientist and the scientist trusts the industrialist."

22
A BETTER IMAGE FOR THE WOLF

Boyce Rensberger

ELY, Minnesota—A crimson pool of blood lay frozen in the snow. Ravens pecked at the last shreds of what had been a deer.

"There they are!" David Mech, wolf biologist, shouted over the roar of a Cessna 180 circling tightly above the frozen lake as he pointed down at the shore. "The wolves."

About a hundred yards from the kill in the middle of the lake, six timber wolves rested drowsily in the cold, bright sun. One jumped up to watch the little plane as it banked sharply.

Dr. L. David Mech, who has studied wolves for eighteen years and is widely acknowledged as this country's leading expert on wolf behavior, marked down on a tracking form the location of the wolves, how many there were, and what they were doing. Then he directed the pilot to head southwest toward the place where another wolf pack had last been seen.

Through such aerial observations, a technique he helped develop, Dr. Mech has gathered much of the evidence that has debunked many of man's oldest myths about the wolf.

Once widely hated and persecuted as a dangerous predator, the wolf today, thanks largely to Dr. Mech's research, is coming to be regarded as an ecologically important member of its wilderness habitats and an animal with a complex and fascinating society. Once feared as dangerous to people, the wolf is now known not only as friendly and sociable within its pack, but as no threat to man. There is no documented instance of a free-living wolf attacking a person in North America.

Dr. Mech (pronounced Meech), now forty years old, began his wolf research in 1959 as a graduate student while observing the two dozen wolves of Isle Royale National Park in Lake Superior. Today, employed by the Endangered Wildlife Research Program of the U.S. Fish and Wildlife Service, he is in charge of a wide-ranging, long-term study of the relatively stable population of 1,000 to 1,200 wolves in northern Minnesota. These animals are the last substantial population of wolves in the United States outside of Alaska.

The greatest concentration of these wolves is in the Superior National Forest near here in the extreme northeastern corner of the state, and the dead of winter, when the beasts can be spotted against the snow, is the best time for studying them.

As the little plane headed toward the next wolf-pack sighting, Dr. Mech put his headphones back on and listened intently for a clicking signal picked up by antennas mounted on the plane's wing struts.

One of the wolves in the pack, like one in the pack sighted near its kill, was wearing a tiny, battery-powered transmitter on a collar around its neck. The signal from this constantly operating device picked up by the antennae on the plane guides Dr. Mech to the wolves.

Over the years Dr. Mech and his assistants, most of them graduate students in wildlife biology at the University of Minnesota, have trapped and radio-tagged about 140 wolves in the Superior National Forest area.

After being captured in a modified leg-hold trap, the wolves are immobilized with drugs, weighed, and identified by sex. Blood samples are taken. Ear tags are clipped on and a collar is fitted. As the drugs wear off, the wolf, transmitting on its own frequency, runs off to rejoin its pack. Because wolf packs are stable social units, the signal from a single radio collar can lead Dr. Mech or his students to the entire pack.

Because the antennae are highly directional, picking up signals from either the right or the left side of the plane, depending on which antenna is used, Dr. Mech can tell where the transmitting wolf is in relationship to the plane. Dr. Mech can switch from one antenna to the other. When the signal is equally strong from the two antennae, the plane is heading directly for the wolves.

Twenty-four of the 140 radio collars are still working, the others either having stopped operating (usually after a year or so) or the wolves have been killed (wolves that venture too near human beings risk being shot or trapped). The twenty-four tagged animals represent nine packs, one newly formed pair that may breed to establish a new pack, and four lone wolves, animals who have left their original packs to wander alone, and perhaps to find a mate and a vacant territory in which they can establish a new pack.

Minutes after leaving the pack with the deer kill, Dr. Mech signaled the pilot to circle above a forested ridge. "They're down there," Dr. Mech shouted. "Can't see 'em. They're probably under the trees."

After logging their position and that of some other packs, Dr. Mech headed back to the airport. Every day during winter and at least weekly during the summer Dr. Mech or his students go up in planes to find the collared wolves. One pack has been tracked for six years.

When the locations for a given wolf pack are plotted on a map, almost all fall within a tightly circumscribed territory abutting the territories of other wolf packs, almost never overlapping them. One pack, for example, has fluctuated from two to nine members over the years, but has always maintained the same territorial boundaries. On a larger scale, wolf densities are usually about one for every ten square miles.

Wolves mark the boundaries with urine and, even when chasing prey, seldom enter alien territory. When they do they risk attack from the resident pack.

Back at the log cabin on a Forest Service compound near Ely that serves as the wolf study's field station, Dr. Mech and about half a dozen students come and go throughout the day, drying out soaked gloves, pouring hot coffee, calibrating radio receivers, exchanging information on the day's sightings. In the refrigerator, vials of wolf blood share space with bottles of beer. Several of the students track wolves from the air and others manage projects on deer, lynxes, moose, snowshoe hares, ravens, and other local fauna. Many of these animals are also wearing radio collars.

By studying wolves and their prey, Dr. Mech and his students hope to discover and understand those elements of their behavior that have evolved as ways of coping with the other species. Anatomical adaptations for attack and defense are well known, but behavioral adaptations are not.

Biologists have long known that wolves have developed certain ways of hunting that maximize their chances of killing deer. Presumably deer, who heretofore have not been intensively studied as one-half of a predator-prey relationship, have evolved defensive behaviors as well. One deer behavior that is under study is the congregating of deer in open meadows, or deer yards, in the winter. In spring and summer deer are dispersed through the forest. Why do deer shift back and forth through the two different systems of social organization?

Somehow, Dr. Mech suspects, the deer's slower metabolism in winter, the difficulty of moving in snow, and the fact that fawns have grown more independent since their birth in the spring make "yarding" a better way to defend against wolves in winter but a poorer way in summer. A four-year study of deer behavior is planned in an attempt to explain this phenomenon.

One morning a student came into the log cabin and reported that wolf No. 2407 and its pack were found well out of their territory. Dr. Mech checked the location on a map. "That's interesting," he remarked. "Those sons of guns, they're trespassing, really striking out on their own." From the map it appeared that they would have had to cross two roads to reach their present location from where they had been the day before.

Later in the day, after a futile attempt to find a radio-collared wolf that someone reported seeing on a road with a trap on its foot, Dr. Mech drove the snow-covered roads that the "trespassing" wolves had crossed to look for scent marks. When wolves cross a road or other physical boundary, they mark the junction with urine. He wanted to collect urine samples frozen into the snow for biochemical analysis, but a snowplow had recently obliterated the marks.

Interactions between Dr. Mech's group and the human population in and around Ely have proven both rewarding and frustrating. Dr. Mech said that although most townspeople were sympathetic to the wolf research and favored the species' protection, a few retained the older antipathy. There is a vigilante group that kills wolves whenever possible and puts the carcasses on other people's doorsteps with notes arguing that wolves destroy deer that should be protected for hunters.

Although wolf hunting and trapping have been illegal in Minnesota since 1965, they continue. Whenever one of the collared wolves is killed, however, some sympathetic trappers notify Dr. Mech by leaving anonymous notes in a certain local bar. Because some of Dr. Mech's colleagues study deer, some townspeople are puzzled that scientists who specialize in species that are seen as "natural enemies" can be friends.

"We have a ways to go in changing peoples' attitudes about these animals," Dr. Mech said. Dr. Mech says he likes to get into the field as often as possible but noted that he has a desk job with the Fish and Wildlife Service in St. Paul. There, with access to libraries and laboratories, he writes scientific papers on wolf behavior, and consults with scientists and conservationists around the world. Last year he spent a month in India training biologists there in radio tracking. He also develops strategies for protecting wolves.

For example, Dr. Mech was heavily involved in the 1974 effort to relocate four Minnesota wolves into Michigan's Upper Peninsula. Within eight months after the wolves were released, all four, two males and two females, had been killed by human beings. Two were shot, one was trapped and then shot, and one was hit by a car.

The experiments did establish that relocated wolves could establish themselves in a new territory and survive. "The problem," Dr. Mech said, "is the human population. Next time we would want to do a more intensive public education effort."

Dr. Mech is a member of the Eastern Timber Wolf Recovery Team appointed by the Fish and Wildlife Service to devise a program for the protection and reestablishment of wolf populations. The group has suggested that wolves be reintroduced to wilderness areas in Michigan, Wisconsin, New York State's Adirondacks, Maine, and the Great Smoky Mountain National Park.

Dr. Mech has found that wolves can double their numbers every year. However, they do not if the area is already full and each pack's territory abuts others on all sides. Lone wolves, unable to establish a territory near their place of origin, disperse to a less desirable habitat and, in many cases, are killed by people.

Thus, Dr. Mech has found, wolf hunting or trapping can continue at a substantial rate on the fringes of prime wolf country without lowering the average wolf population. In Minnesota, for example, the Wolf Recovery Team has recommended that controlled wolf killing be permitted in a buffer zone around the wolf's 10,000-square-mile prime range, which would remain totally protected. Limited wolf hunting or trapping, the group believes, is necessary to minimize the loss of livestock to wolves and to increase the base of local citizen support for conservation, without which wolves might not survive at all.

23

RATS: LIVING MONITORS OF RADIATION

John Noble Wilford

ENEWETAK, Marshall Islands—About once a year Bill Jackson travels a third of the way around the world to go hunting. He hunts rats.

He could undoubtedly stay home in Bowling Green, Ohio, and find plenty of rats. They would be Norway rats, the misnamed species all too common to the slums, sewers, and dumps of the temperate zone. But he has his mind set on tropical-island rats, particularly those whose ancestors somehow survived the nuclear tests that wracked this central Pacific atoll between 1948 and 1958.

William B. Jackson, professor of biology and director of environmental studies at Bowling Green State University, is an expert in rodent ecology. He is a "rat person," as he is accustomed to being called, and since 1964 he has been coming here nearly every year to trap and study the rats of Enewetak.

Much of his work is aimed at learning the habits and behavior of island rats, as compared with city and laboratory rats. But, for the same reason that laboratory mice and rats are used to test new drugs, food additives, and other consumer products, he is looking at the Enewetak rats as living monitors of any residual radiation on the islands of the atoll most affected by the nuclear blasts.

The investigations by Dr. Jackson and other scientists

have found that the Enewetak rats survived the nuclear tests with little apparent effect. The organs of some rats contain slightly larger-than-normal amounts of radioactive cesium, which they got from eating contaminated plants. They have suffered no apparent genetic damage, although little study has been done in this area since the immediate posttesting period.

The scientists have found the island rats to be relatively tame, having no predators to fear and having had little contact with man for several decades. Compared to city rats, those on Enewetak are clean animals, having little contact with garbage, filth, or disease. They will eat just about anything, but primarily their diet seems to be berries and sand burrs.

Explaining his work one day recently, Dr. Jackson said: "This is an umbrella effort to determine the general ecology of rodents at Enewetak—their food habits, reproduction, behavior, population dynamics [numbers and longevity], and the role of the rat in the entire ecosystem. The rat is pretty much the top consumer in the food chain here, besides people, and on most of the islands he has no predators.

"Using this knowledge as a base, we have looked at the role of the rat as a biological monitor for radioactive material. We'd like to do even more such studies, seeing how any radioactive material still in the soil gets into plants and up the food chain to mammals. We think we can tell a lot with rats that can't be told in any other way."

With Dr. Jackson on the 1977 hunt was Stephen H. Vessey, an associate professor of biology at Bowling Green. Dr. Vessey's objectives are more specific: to test "in different species and a different environment" a theory that the stresses of crowding, as reflected in fighting and physiological changes, serve to regulate rat populations.

After lunch one day, the two scientists set out across the

lagoon on an overnight expedition to Enjebi, one of nearly a dozen of the forty tiny islands in the atoll where they collected rats. Phillip Lamberson, manager of the Mid-Pacific Marine Laboratory at Enewetak, piloted the Boston Whaler on the two-hour, twenty-mile voyage.

Mr. Lamberson brought the boat up to the coral beach in front of a house trailer maintained on Enjebi by the marine laboratory. Everyone wades ashore, hauling the gear—bedding, gunnysacks filled with snap traps, and two ice chests. One chest was for food, the other for beer, and both would be used to bring back the dead rats.

Hunting rats is night work. And so, late in the afternoon, Dr. Vessey went over to the ocean side of the island to set about seventy-five traps in the brush along an abandoned airstrip. Dr. Jackson shouldered his sack of seventy-five traps and headed down a sandy road bordered by dense scaevola scrub. Rats feast on the white scaevola berry, as well as on sand burrs, coconuts, insects, lizards, fish washed ashore, and even dead rats.

After pacing off about twenty-five feet, Dr. Jackson pulled a trap out of his sack and a small piece of fresh coconut, the bait, from his pocket. He cocked the mechanism and placed the trap off in the undergrowth. Then he dug a line in the road with his shoe to mark the place, and set off down the road, repeating the procedure every twenty-five feet.

Dr. Jackson explained that there were two kinds of rats on Enewetak, the roof rat and the Polynesian rat. The roof rat, the larger type, is common throughout the tropics and probably arrived at Enewetak during World War II. The Polynesian rat presumably came in the boats of the first settlers centuries ago.

A study in 1957, involving the capture and marking and recapture of rats, indicated that there were about 10,000 roof rats on Enjebi. This meant, Dr. Jackson explained, that some rats—possibly as few as fifty—must have survived the thermonuclear test of 1952 that seared Enjebi with radiation and heat, blasted it with shock waves, and scoured it with tidal waves.

The survivors must have been deep in their holes or under some of the test bunkers. They not only survived but multiplied many times over. The Polynesian rats, not being burrowing animals, were apparently eradicated on Enjebi.

"It seems kind of remarkable that any survived, but rats are remarkable animals," Dr. Jackson said. "They're adaptable. They change and live in a tremendous number of habitats. They'll always do very well."

When it was suggested that, in the event of a worldwide nuclear holocaust, rodents might inherit the Earth, Dr. Jackson paused between trap settings and nodded his head. "There's something to that," he said.

Life has returned to Enjebi. Scaevola bushes and kirin trees abound, as well as tangled vines, some palms, hermit crabs, and many birds. Enjebi is considered safe to visit but may not be safe for permanent habitation for another thirty years. It was once home for half of the Enewetak people.

As soon as Dr. Jackson and Dr. Vessey returned to the camp, Mr. Lamberson put the steaks on the charcoal grill. Steaks and beer and a beautiful sunset.

Well after dark the scientists went out to check the traps. If they left the trapped animals out there too long, crabs or other rats might get there first. Dr. Vessey recalled once having had a tug-of-war with a hermit crab for a trapped rat.

Dr. Vessey came back first, disappointed. He had only five rats, one of which he had caught himself, stunning it with the light from his headlamp and then stepping on its head.

Dr. Jackson returned with fourteen rats, and they all were laid out on the floor of the trailer porch for a cursory examination. After checking the sex of the rats, Dr. Vessey sighed, "Well, there goes that theory."

He had an idea that the males, particularly the dominant males, would be the first to venture out each night. But there were more females than males in their catch. On some other islands, however, the males did seem to be the first out.

Everyone slept on the open porch, lulled by the sound of the surf and the cricketlike music of the gecko, a small lizard. The scientists made two more trap runs, at midnight and at dawn.

Forty-three rats were on ice when the party loaded the Whaler and returned to Enewetak. Dr. Jackson and Dr. Vessey were eager to begin the autopsies before decomposition set in.

Working at a table in the marine laboratory, Dr. Jackson logged in each specimen, noting the sex and size. All were adult roof rats, about six to eight inches long, with tails as long as or longer than the bodies.

A few of the specimens were decapitated so that the heads could be shipped back to Bowling Green for more detailed examination. Other scientists have found that the palates of rats in the Galapagos Islands have markings that differ from island to island. Dr. Jackson wants to see if the same holds true for rats from different islands of Enewetak.

Dr. Jackson then made abdomen incisions and examined through a microscope the contents of the stomachs to see what the rats were eating and to note the extent of parasitic life, which was considerable. Dr. Vessey extracted and labeled the intestines for further study as well as some of the organs, particularly the adrenal glands.

The adrenal of a rat is half the size of a small pea, but it could be the key to Dr. Vessey's studies of stress as a factor in the regulation of rat populations. Dr. Vessey placed each adrenal in a separate labeled vial of formalin, a preservative that hardens the gland so that it can be more easily cleaned and weighed the next day.

Then, with an electric shaver and a commercial hair-removing lotion, he cleaned the backs of each specimen from the shoulder to the tail. He was searching for wounds and scars of past battles. Scars around the shoulder, Dr. Vessey said, could indicate a dominant rat, one that had been in head-to-head combat with an equal. Scars down the back and around the tail could indicate a rat that had been chased and bullied, a lesser creature in the rat society.

After the field work is completed, Dr. Vessey intends to compare the size and condition of the adrenals of dominant rats with those of scarred rats. An enlarged adrenal, he said, is a "prime indicator of chronic stress." Moreover, laboratory studies have shown that such stress delays maturation, inhibits sexual activity, and may even weaken the rat's resistance to parasites and disease.

If a correlation between enlarged adrenals and fighting scars can be confirmed, Dr. Vessey said, this should be evidence to support the theory that fighting among rats, over food or space or both, probably causes physiological changes that, in turn, tend to regulate the size of an area's population. This idea is called the "general adaptation syndrome," or self-regulation.

Although the concept is not completely accepted in the scientific community, it has many adherents because of the so-called classic studies of rats in the slums of Baltimore by the U.S. Public Health Service in the 1940s and 1950s and the more recent laboratory studies of rats and mice by Dr. John B. Calhoun of the National Institutes of Mental Health.

Dr. Calhoun attempted to demonstrate how the animals use space and how crowding affects their behavior, extrapolating from that the more controversial idea that overcrowding among humans may likewise account for violent or antisocial behavior. According to Dr. Vessey, there have been "very few good studies of rats in their natural environment since the Baltimore study" and it is "important in science to take a significant idea and test it again and see if it might apply elsewhere."

As Dr. Vessey and Dr. Jackson worked on Enjebi and in the laboratory, they spoke of themselves and how they came to be "rat persons." It provided an insight into the sociology of scientists, the things that influence their careers and the types of research they embark upon.

Dr. Jackson was born fifty years ago in Milwaukee and attended the University of Wisconsin. While he was a graduate student there, one of his professors had just come from Johns Hopkins University, where he had participated in the Baltimore rat study. He encouraged Dr. Jackson to do his doctoral studies at Johns Hopkins, where he, too, became involved in the rat study.

Dr. Vessey, now thirty-eight years old, was born in Stamford, Connecticut, and grew up in White Plains, New York, the son of a Unitarian minister. After undergraduate studies at Swarthmore, he attended graduate school at Pennsylvania State University. One of his most influential biology professors there, Dr. David E. Davis, was a specialist in animal behavior who had known Dr. Jackson at Johns Hopkins.

Through such coincidences, Dr. Vessey became a member of the faculty of Bowling Green. And though he insists he is "not strictly a rat person," but more a specialist in ecology and animal behavior, Steve Vessey now finds himself traveling, with Bill Jackson, a third of the way around the world to hunt rats at Enewetak.

A year later, Dr. Vessey gave the following report on his studies of the relationship between fighting and physiological changes in rats captured at Enewetak: "Data thus far analyzed support the hypothesis for females, but not for males. There was a significant positive relationship between the number of back scars in female Polynesian and roof rats and their adrenal weights. In other words, rats that were victims of aggression were under more stress and therefore had heavier adrenal glands than those that were attacked less often.

"A surprising result was that for both species, on six of nine islands, females had more scars than males. Usually we think of male-male fighting for status, territory, or access to females. Females are thought to fight little, except when defending the nest and young. Our data suggest that females are involved in more aggression than males and may be the key to understanding population regulation in rats."

24

THE COUNTLESS MYSTERIES OF PEATLAND

Jane E. Brody

BIG BOG, Minnesota—Like the meadow in Andrew Wyeth's painting *Christina's World,* the sprawling field of tall grass shimmering in the gentle breeze seemed a fine place to spread out a picnic—until you stepped into it and sank to your ankles in cold water.

For it wasn't a meadow at all and the rippling vegetation wasn't grass. Rather, it was a plant called sedge, one of hundreds of species that live with perennially wet feet in a soggy peatland that covers some 300 square miles of northern Minnesota.

Because of an ever-tightening energy supply, the many layers of peat in Big Bog, as it is called on some maps, are attracting interest as a possible source of natural gas. But to Dr. Eville Gorham, ecologist from the University of Minnesota who frequently treks into Big Bog with a dozen or so students, the waterlogged terrain is a source of countless mysteries just waiting to be unraveled.

"Somehow, I find it hard to look out over this and see nothing but BTUs," Dr. Gorham remarked as he gestured across the expanse of sedge, his brown eyes reflecting the tranquility of the waving strands of green that were almost silvery in the late summer sun.

Dr. Gorham is one of a very few modern scientists ("you can count us on your fingers and toes") who study the ecology of peat bogs, those places commonly called swamps that are not wet enough to boat through or dry enough to walk through without getting wet.

Dr. Gorham's students say that peatland ecology is a very primitive science for which there is thus far no text. "We read scientific papers and long monographs, take a lot of field trips, and do our own studies," explained one young woman from Bemidji, Minnesota, who said she was taking Dr. Gorham's course because, although she grew up in the midst of peatlands, she knew nothing about them. "To most of us, a bog is a swamp is a marsh," she observed.

The lack of study and appreciation of peatlands is no mystery to Dr. Gorham. "Most people either keep their feet dry or go out in boats, but they don't like to muck around where it's squooshy. Also, they don't like the mosquitoes and they're afraid of falling in."

Indeed, in some boggy areas you can step onto what appears to be a safe patch of moss raised above the water and quickly sink clear up to your hips. The particular area Dr. Gorham and his students visited this year was more like a waterbed—you could actually bounce on it and the fibrous mat of sedge roots some eight inches thick would quake, but you wouldn't fall through.

The sedge "fen"—as the Europeans call this type of peatland—is fed by waters that percolate through mineral soils, as well as by nutrient-rich runoff. But according to Dr. Gorham's definition, a true bog receives its nutrients almost solely from the atmosphere—from rain, dust, industrial pollution, or sea spray—and supports a wide range of vegetation, from sphagnum (peat moss) and lichens to tiny birch and tall black spruce.

A swamp—more a popular than a scientific term—is a forested wetland that can be either a bog or a fen, usually with water standing or flowing through or over it. A marsh, on the other hand, has few or no woody plants. Rather, grasses and reeds grow in its silty soils.

Most peatlands develop in flat areas—the region of Big Bog was flattened by glaciers—where the climate is cool and damp and the water supply is subject to damming, often by beavers.

The unschooled observer may not see anything special about peat bogs. They are neither as eerie or exotic as the great swamps, nor do they teem with apparent wildlife.

Yet to a scientist like Eville Gorham, their distinctive ecology makes them a fertile area for deciphering the delicately balanced relationships between plants, animals, and their waterlogged environment. They also possess some rare properties that have proved historically as well as biologically fascinating.

In May 1950, a professor named P. V. Glob was called to examine a very well-preserved body that had just been found in Tollund Fen, a Danish peat bog. The workers who unearthed the body from eight feet of peat thought they had stumbled upon a recent murder victim.

Instead, Professor Glob wrote in his book, *The Bog People,* the body was that of a man who lived 2,000 years ago, a relic of the Iron Age preserved as if impregnated with a miraculous embalming fluid by the acidic, antiseptic, oxygen-depleted waters of the bog. More than 700 such ancient bodies have been dug out of the bogs of northern Europe, all remarkably well preserved.

Whether any bodies would be found if and when the peat is harvested from northern Minnesota is anybody's guess. But what can be found right now are layers of sphagnum moss so well preserved that moss two or three meters down can be identified by species, Dr. Gorham said.

Sphagnum is the signature plant of a bog. Its large water-holding cells give it a spongelike quality that allows it to absorb 200 times its own weight in water. Although most commonly used to condition the soil in gardens and houseplants, sphagnum has also served as litter for house pets, baby diapers, and, in World War I, as wound dressings, the scientist said.

When compacted into peat and dried, sphagnum has twice the heating capacity of an equivalent amount of wood. Peat is formed in cool, waterlogged places where the lack of oxygen inhibits bacterial decay. Sphagnum further resists decay because it is highly acidic and contains nitrogen in a form not usable by bacteria.

Beyond the reeds and cattails along the roadside ditch, the sedge fen seemed at first to be a monotonous stretch of grasslike greenery. But as Dr. Gorham slushed through the sedge, he exposed the incredible variety of life hidden in the bog.

Parting the sedge, he unveiled a clump of bladderworts, small flowering plants with tiny pouches that sit in the water and catch and digest microscopic animals that happen past. Then he spotted an insectivorous plant, the sundew, a delicate rare species with sticky droplets that glisten in the sun.

Scattered here and there were patches of mosses, marsh fern, irises, low willow, and buckbean, which has airholes that carry oxygen to the waterlogged roots to keep them alive.

Deeper into the sedge fen, cushions of sphagnum began to appear, creating spongy hummocks raised above the water surface where plants that can't live in water can get a "roothold."

The highly poisonous water hemlock, leatherleaf, cranberry, sweet gale (the crushed leaves have a cardemonlike scent), bog rosemary, and bog birch (a low shrub) were among the many species of plants living on the sphagnum hummocks.

Finding one large hummock honeycombed with the runs of small mammals, Dr. Gorham remarked, "This is the animals' island in the sea." He said large populations of weasels, voles, mice, and rabbits, among others, live in the wetter areas (a marsh hawk hovering overhead was looking for one for lunch), and moose and deer live in the forested fens. But, he added, no systematic study has yet been made of the animal life in peatlands.

Many of the hummocks also housed pitcher plants, another rare species with leaves like rams' horns that catch and hold rainwater. Insects that crawl into the pitchers are trapped by downward-pointing hairs and then digested by a plant enzyme. At least one species of mosquito has evolved resistance to the enzyme and instead uses the water in the pitchers as breeding pools.

Still further into the bog, larger and drier hummocks were populated with small, shallow-rooted trees—tamaracks and black spruce, with labrador tea, lichens, and lignonberries at their feet. Throughout Big Bog, teardrop-shaped "tree islands," their long axis parallel to the direction of water flow, dot the flat landscape.

Now it was time for the real work. Dr. Gorham and his students sank metal corers deep into the hummocks to study their evolution. They circumscribed areas to count the plant species within. They took the temperature of the soil and light levels at the surface, measured tiny changes in the soil and water levels, and collected water samples to analyze their chemical contents.

In previous studies, carbon dating of the bottom layer of peat eighty-five inches down showed it to be 4,360 years old, indicating that about 2 inches of peat accumulated each century, a much slower rate than the present peat build-up.

Still, Dr. Gorham is concerned about the prospects of mining Big Bog's peat. "It will mean taking hundreds of thousands of acres, the consequences of which are hard to predict," he said.

He continued, "We ought to preserve major sections of these wetlands, partly for esthetic reasons and partly for scientific interest. There's a lot to be discovered. You never know what these areas may be able to tell us that nothing else can. What if man had gotten rid of the Galapagos finches and put in pigeons before Darwin had a chance to develop his theory of the evolution of species?"

VI
Science and Health

BACTERIOLOGY, NUTRITION, VIROLOGY

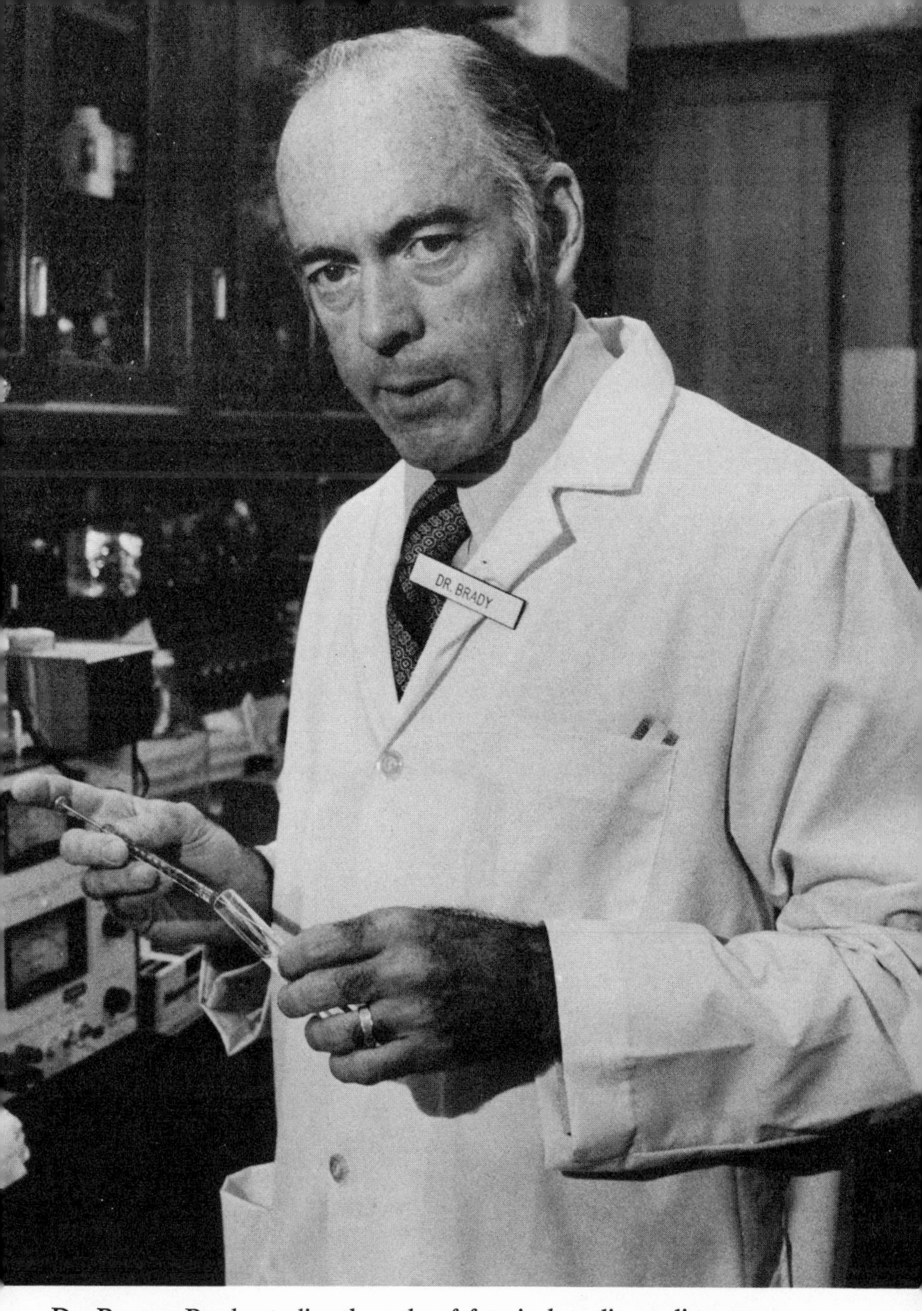

Dr. Roscoe Brady studies the role of fats in hereditary diseases for life-saving clues.

Dr. Edwin Kilbourne testing swine-influenza virus in his laboratory at Mt. Sinai Medical Center.

Doctors Joseph E. McDade, left, and Charles C. Shepard discussing discovery of the cause of Legionnaires' disease.

Dr. Stanley Falkow in lab in Seattle, Wash., working on test of resistance by organisms to antibiotics.

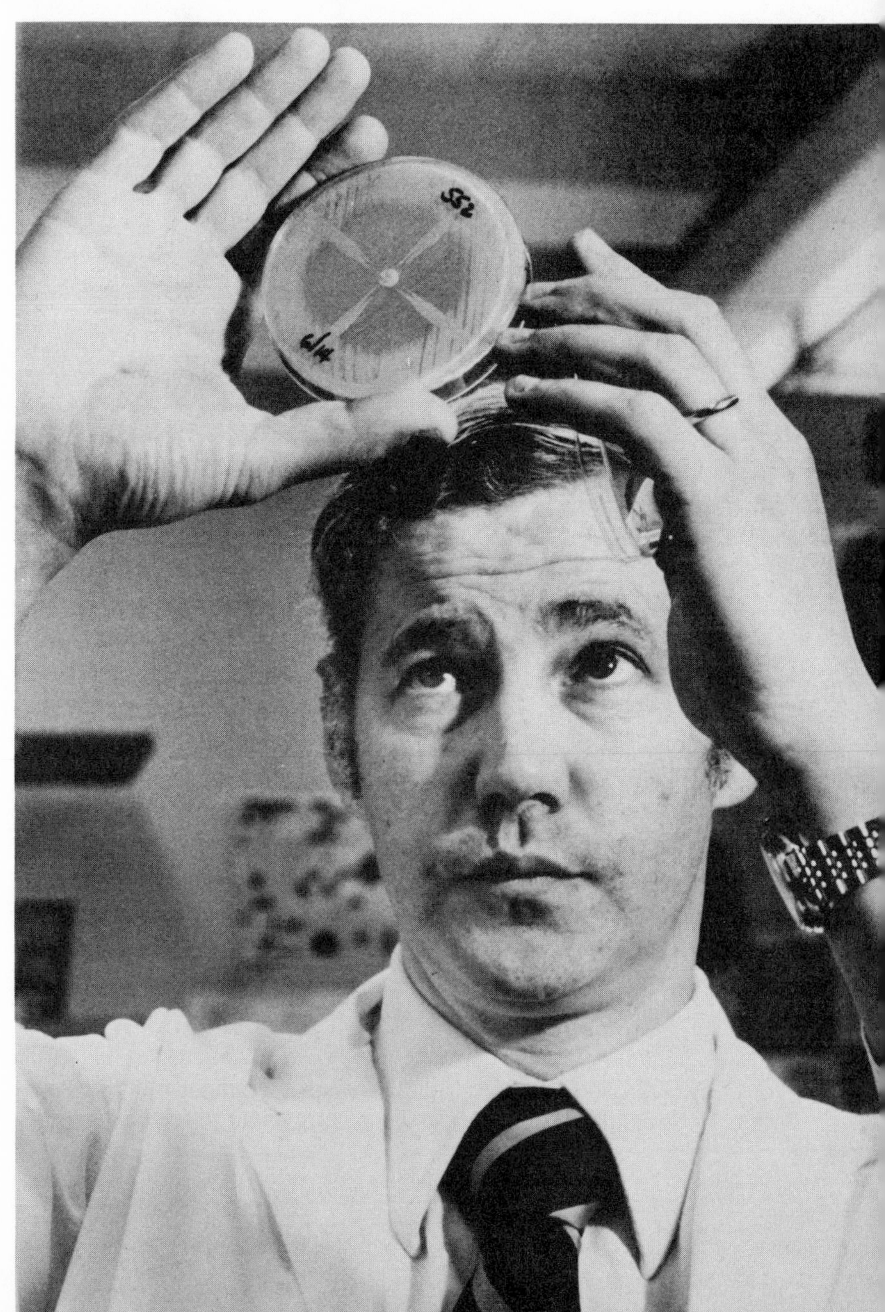

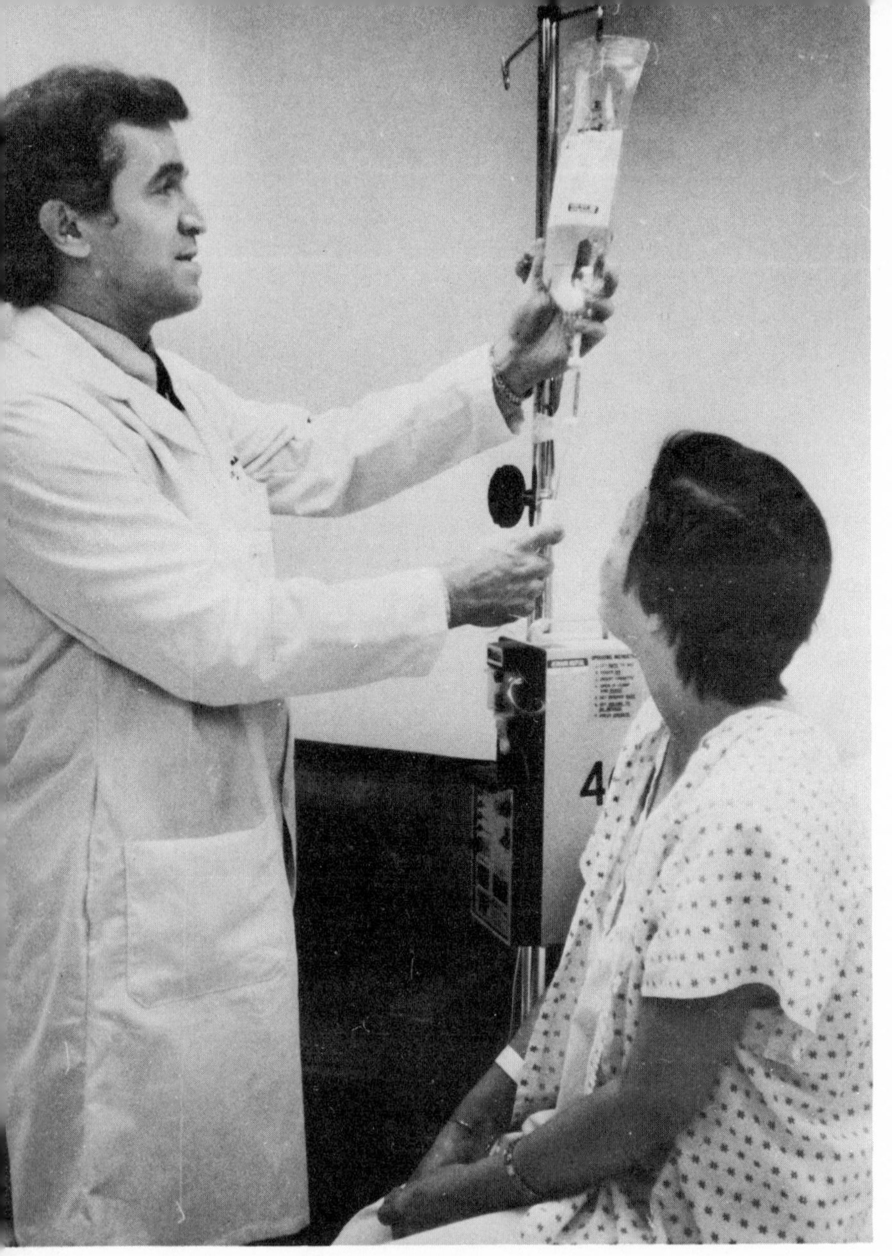

Dr. Stanley J. Dudrick with a 25-year-old patient who has severe ileitis and lives almost entirely on intravenous nutrition.

25

FAT AND LIFE-SAVING CLUES

Harold M. Schmeck, Jr.

WASHINGTON, D.C.—The death of a patient is a serious blow to any doctor. When the "doctor" is a young medical student, the shock can be even deeper and more enduring.

During the 1940s medical students did a lot of almost everything related to patient care to help cope with wartime shortages of doctors and nurses. This often brought the students very close to patients and their problems. That is how a young man named Roscoe Brady came face to face with two tragedies that are still fresh in his mind several decades later.

They were the deaths of a young mother and a middle-aged man, both of heart disease. The deaths led the medical student to pursue a career in research devoted to a subject at the heart of most heart disease—the body's use and disposal of fatty materials such as cholesterol. As a result, a whole new frontier of knowledge has been opened in the field of hereditary diseases.

Some day, as a consequence of this, injections of an enzyme unknown in the 1940s may save the life of a child otherwise doomed by a defect in heredity that was hardly guessed at in those years.

Although many specialists see formidable obstacles still ahead in attempts to treat patients by enzyme replacement, Dr. Brady is hopeful of bringing such a treatment to the point of practical value for some patients. When and if this is accomplished a new era will be opened in the treatment of hereditary disorders that have been under study for more than a decade at the National Institutes of Health. They are called lipid storage diseases.

They are rare, with names unfamiliar to most people—Gaucher's disease, Fabry's disease, metachromatic leukodystrophy, fucosidosis, Tay-Sachs disease. They are incurable and at any one time several thousand American families are burdened with the tragedy resulting from one of these conditions.

The symptoms vary from disease to disease. Several cause mental retardation. One causes kidney failure, another causes blood-clotting problems and erosion of the bones so that they become fragile and easily broken. One causes blindness.

The link between these devastating mistakes of heredity and the two heart-disease deaths thirty years ago is the word lipid.

Lipids are fatty substances necessary to life. An excess of some of them, notably cholesterol, has long been considered a key factor in heart disease.

Other lipids, breakdown products of normal body constituents, accumulate in the bodies of the victims of lipid storage diseases. The effects of this accumulation are devastating. In each case, the cause is the lack of a single enzyme among the many thousands in the human body: a different one for each disease.

Enzymes are chemicals, produced in the body, that act as catalysts to aid different chemical processes on which the body depends. When an enzyme is either missing or defective in function an important chemical process is deranged.

The lipid storage diseases are examples of this subtle but catastrophic derangement. They are much less mysterious than they were three decades ago, partly because Dr. Brady knew a twenty-six-year-old mother of two who died of rheumatic heart disease late in her third pregnancy, and a man in his fifties who died during surgery to restore some of the blood circulation to his heart, which had been severely afflicted with arteriosclerosis.

While it was these two heart-disease deaths of patients under his care that led Dr. Brady into his research, only the latter case was related to excess of fatty materials. Lipid abnormalities were then—and still are—the central puzzle of arteriosclerosis. The process of arteriosclerosis, in which arterial blood flow becomes progressively blocked, is, in turn, the central puzzle of most heart disease.

"At that time I really decided we needed to know something more about arteriosclerosis," said Dr. Brady. Virtually nothing was known then about the way the body put cholesterol together and took it apart or why the process went awry. It seemed a good focus for a career in medical research.

Dr. Brady, a tall, spare man who plays squash, tennis, and the piano in what free time he has, also made a pact with himself early in his career—either accomplish something significant by the age of thirty or quit research and practice medicine.

From that time on the intricate puzzle of lipid build-up and breakdown was an irresistible attraction to which he always returned. But over the years his interest was sometimes deflected and his progress often seemed agonizingly slow.

Early in his research career, Dr. Brady spent a whole year trying to synthesize a compound that he believed to be a key to the lipid problem, but he failed in the synthesis. Furthermore, it proved to be a false lead.

Later, however, he worked out the details of a previously unknown enzyme system that was important in the natural production of lipids. It was a significant research accomplishment, satisfying his pact with himself. But then the

Korean War erupted and he was called by the Navy.

During the war he served at the Naval Medical Center across the street from the National Institutes of Health. After his service he was hired by one of the component institutes of the N.I.H.—the National Institute of Neurological Diseases and Blindness.

Dr. Seymour Kety, an officer of the institutes and one of the pioneers in study of the chemistry of the nervous system, wanted someone to work on lipids. It was an important area in his field too, because the fatty materials are the major constituent of myelin, the substance that sheathes and insulates important nerves.

This led Dr. Brady to research on the body's system for making sphingosine, a key ingredient in myelin; and sphingosine brought him directly to the lipid storage diseases.

The disorders are all characterized by abnormal accumulations of fatty materials called sphingolipids, of which sphingosine is a part.

Since the middle 1960s, Dr. Brady and his colleagues have made many important discoveries concerning the chemistry of nine different hereditary lipid storage diseases. They have developed important diagnostic tests for three of these—Gaucher's disease, Niemann-Pick disease, and metachromatic leukodystrophy—and have begun experimental treatment of Gaucher's and two others, Fabry's disease and Tay-Sachs disease.

Of most immediate importance, the research has led to major advances in detection of persons who carry the genetic traits that could put some of their children at risk. In most of the disorders there is a risk only if both parents carry the trait. In that event the odds are one in four that any child will be afflicted.

In terms of science, the work that began more than a decade ago with sphingosine in the central nervous system has led to understanding of a whole group of genetic disorders. In time this may lead to effective treatment for at least some of them.

An article a few years ago in a periodical of the American Chemical Society described the scientist's work as the discovery of a "biochemical Rosetta stone."

Like the original Rosetta stone, which gave archeologists the key to Egyptian hieroglyphics, the work with lipids has provided a key to many previously baffling observations. It shows why the fatty materials accumulate so disastrously in the lipid storage diseases and—in theory at least—what can be done about it.

In finding his "Rosetta stone," Dr. Brady and his coworkers used many of the sophisticated weapons of modern chemistry: whirring centrifuges that look like washing machines from the outside but, inside, spin samples of blood or tissue at 50,000 revolutions a minute to separate the substances scientists want to isolate from the sample's "chaff"; scintillation counters that record and count every atomic eruption in a sample of material "tagged" with artificial radiation so that its chemical transformations can be traced; tall columns of special chemicals designed to separate one enzyme from a complex liquid mixture.

The first of these problems that Dr. Brady's group attacked was the disorder called Gaucher's disease, discovered by a French physician named Phillippe C. E. Gaucher in 1882.

Fatty material accumulates in the liver, spleen, and bone marrow of the victims of the disease. Blood clotting is hampered. Bones erode and are easily broken. The patient may suffer agonizing swelling of joints. Those most seriously affected do not live beyond childhood. In these cases lipids also accumulate in the brain, causing severe mental retardation and death.

It is a rare and tragic disease and a difficult problem for medical research. There seemed to Dr. Brady to be three possibilities: the patient's body might be making abnormal material, it might be making too much of something normal, or it might be unable to dispose of something that was normal and made only in normal amounts.

There was no way of guessing which was true. Each hypothesis had to be tested by research.

The first two parts of the problem were solved with the aid of slices of spleen growing in laboratory flasks.

The spleens of patients with Gaucher's disease sometimes have to be removed to cope with disease-caused abnormalities of the blood. Living in laboratory tissue cultures, these slices of human tissue kept on making the fatty materials that accumulate in Gaucher's disease. Studying this disembodied production, the scientists learned first that the material was normal—not significantly different from the material produced by a normal person's spleen. Many more months of study showed that the production rate was also normal.

The possibility that the accumulated material was itself abnormal was ruled out in the mid-1950s, but it took many months to do it. The second possibility—too much production of the fatty material by the body—was ruled out in 1959. At that point, Dr. Brady and his colleague, Dr. Eberhard Trams, offered the hypothesis that Gaucher's disease was caused by lack of a specific enzyme to break down the substance that was accumulating.

That hypothesis put their scientific reputations on the line. They had to go back to the laboratory and prove the point.

It took about five years of sophisticated chemical detective work to describe an enzyme that the scientists suspected to exist in the normal person. But they still had not isolated or named it.

Their next task was to show that this still-hypothetical

enzyme was scanty or absent in Gaucher's patients' tissues, but present in other persons. They found that the characteristic fatty material of Gaucher's disease would have to be made from scratch in the laboratory and some of its parts tagged with radioactivity so that the scientists could catch an enzyme in the act, so to speak, of breaking the material apart.

The first need was a way to tag the complex lipid with radioactivity so that its final fate and breakdown in the body could be traced and understood. The substance is called a glucocerebroside. At that time no one had ever made it in the laboratory. No one had succeeded in tagging it. The N.I.H. team had to do both.

Dr. Brady enlisted the aid of an Israeli chemist, Dr. David Shapiro, who had done impressive work with similar compounds. The institute sponsored a visit to the United States by Dr. Shapiro. In three months he, Dr. Brady, and Dr. Julian Kanfer produced the tagged lipid and used it to prove that Gaucher's disease was caused by the patient's congenital lack of an enzyme that normally breaks the glucose away from the cerebroside.

The enzyme has been given the name glucocerebrosidase. Today it is used as the basis of a diagnostic test to identify cases of Gaucher's disease and to estimate their severity. Recently three patients have been given injections of the enzyme in the hope that it will clear out the accumulation of disease-producing fatty material in their bodies.

Dr. Brady and his colleagues, Dr. Peter G. Pentchev and Dr. Andrew E. Gal, believe this form of treatment may have an important future. It appears that the enzyme may need to be given only once a year or so to remove accumulated stores of lipid and to keep the substance from accumulating again.

The scientists have been extracting the enzyme from placentas collected after the birth of babies.

At present it takes as much as a year to collect and purify

enough of the enzyme for one treatment.

One of the key questions concerning the usefulness of the enzyme treatment remains unanswered for a frustratingly simple reason. Much of the disease damage is done by fatty accumulations in the bone marrow. Does the enzyme clear out these deposits too? In order to find out it is necessary to do a bone biopsy—puncture the bone with a small needle and remove a little tissue for study.

Unexpectedly, it proved impossible to get a satisfactory bone biopsy sample from a patient who had received a massive dose of the enzyme. Months will pass before the same test can be tried again.

Today nine hereditary lipid storage diseases are known. In each disorder there is a specific—seemingly almost trivial—genetic defect involving one enzyme among many thousands. It is a different enzyme in each disease. In each case the lack of the single enzyme has devastating consequences.

But, at the same time, each disease is giving scientists a key to the manner of breakdown of a different fatty material in the human body.

Overall, the scientists investigating the tragic puzzle of the lipid storage diseases describe themselves as cautiously optimistic. They are gratified to see pieces of the puzzle fall into place, but they know much more remains to be learned.

To date they have not saved a life or prevented brain damage in any of the victims. It may be years before they do so. But they have learned enough about the diseases to believe that their quest is by no means hopeless.

26
THE RACE FOR A SWINE FLU VACCINE

Harold M. Schmeck, Jr.

NEW YORK—The race for a new flu vaccine began in February with a telephone call on Friday the 13th.

It was Dr. Martin Goldfield of the New Jersey State Health Department calling Dr. Edwin Kilbourne of Mt. Sinai Medical Center in Manhattan. The message: four samples of flu virus were on their way to New York. They were unusual; nothing at all like the virus called A/Victoria that was the main cause of flu that season.

It meant that another virus against which most Americans had no immunity was abroad in the population and might be gaining momentum for nationwide epidemics.

Clearly, Dr. Kilbourne, chairman of the microbiology department of the Mt. Sinai School of Medicine, would want to grow the virus in his laboratory as soon as possible. He had done this with every major flu variant that had appeared in the United States during the past decade. Each time, he had tried to remodel the virus for rapid growth, making it more useful for vaccine production.

It was a strategy for putting the human race one step ahead of the ever-changing flu viruses. At least four of Dr. Kilbourne's remodeled viruses—called recombinants—have been used in vaccines since 1968.

When the telephone conversation took place in mid-February 1976, no one anywhere was thinking of a huge program to vaccinate the entire United States population against flu. That was not decided upon, or announced by President Gerald Ford, until late March.

But the experiments about to begin in Dr. Kilbourne's laboratory would be important to that program.

The new viruses had been found in recruits at Fort Dix, New Jersey, where there had been a lot of flu in recent weeks.

Dr. Goldfield told Dr. Kilbourne that the new viruses seemed to grow poorly. They would have to be "persuaded" to grow better if drug companies were to use them to make vaccine for the next winter's flu season. The manufacturers were already well along in preparing vaccine against A/Victoria. There would be no chance of also preparing for the Fort Dix virus, unless the vaccine-makers got something useful within a very few weeks. But in mid-February, time was already running out.

Dr. Kilbourne and his laboratory technician, Barbara Pokorny, expected to find the samples waiting for them when they arrived at the medical center that Monday morning, but the package was not there, nor could they find it anywhere at Mt. Sinai.

By that time, they had a new reason for urgency. The viruses had also been sent to the federal Public Health Service's Center for Disease Control in Atlanta. Experts there identified them as viruses of the type that cause flulike illness in pigs—swine influenza virus. This meant they might be related to the virus that caused the greatest influenza disaster in modern history—the worldwide epidemic of 1918–19.

"We were going frantic," Dr. Kilbourne recalls of the search for the missing viruses that Monday. Everyone had overlooked the fact that this was the Washington's Birthday weekend. The virus samples were still somewhere in the mail between Trenton and New York.

On Tuesday, February 17, the samples arrived—one package containing four little glass vials with black screw tops. In each was a colorless fluid containing at least 100,000 infectious virus particles. The viruses from which each sample originated had come from the throat of a soldier at Fort Dix.

The virus had been found in five men, but one of these, an eighteen-year-old recruit, had collapsed on a training march and was dead of influenza-pneumonia by the time medical corpsmen got him to the base hospital. Even though any flu vaccine would be made of inactivated, noninfectious virus, it was thought unwise to use virus from a fatal case.

The four samples had arrived and were enough to work with, but there was another roadblock—no eggs. The virus had to be grown in fertilized chicken eggs, but the holiday had delayed delivery.

The eggs arrived the next day and the experiments began.

All that week and the next, while the nation's influenza experts were debating the meaning of the Fort Dix incident, the virus that had caused all the excitement was adapting to a new world.

The unimaginably small particles, shaped like fuzzy balls, were growing and multiplying in fifty-four carefully labeled eggs in a warm, dimly lit room in Mt. Sinai's Annenberg Building sixteen floors above Central Park.

"This is really animal husbandry in a sense," Dr. Kilbourne explained to a visitor.

Each sample of the new virus was inoculated into eggs together with another flu virus called PR-8, known for its excellent growth characteristics. PR-8, a museum piece among human flu viruses, came originally from samples collected in the early 1930s. Its genes have been mapped and all its characteristics are thoroughly known.

Flu viruses have the helpful tendency to exchange bits of their hereditary information—their genes—when two different strains grow together in the same place. The research strategy was to invite the new virus from New Jersey to pick up the fast-growing habits of PR-8 while still retaining its own recognizable identity.

So samples of the new virus and the old were put together in the same egg—with a serum designed to keep PR-8 from multiplying.

Dr. Kilbourne and his laboratory technician would be looking for viruses that had picked up the talent for fast growth from PR-8 without having lost the traits that identified them as the swine flu influenza type.

At twenty hours, forty hours, sixty hours, Miss Pokorny harvested the fluid from the eggs and tested it to see how much virus was present, and what kind of virus it was. Then she put samples from this new testing into new eggs for a further period of growth. It was like a relay race in which each virus generation passed on the baton to the next generation while the scientists watched to see which sample was winning the contest in fecundity.

The complex routine of virus "animal husbandry" was repeated time after time—harvesting, analyzing, separating, diluting, and growing again. Each step and the history of each egg was carefully recorded.

The work went on without a break, in isolation and almost in secret. Dr. Kilbourne had told no one but Miss Pokorny what the new viruses were. Laboratory safety measures were extraordinary even for flu virus. He wanted no risk of infecting anyone.

"I wouldn't let anybody in the laboratory," Miss Pokorny recalled. "They really thought I had flipped out."

As the work progressed, samples that seemed to be growing best were subdivided by diluting their fluid, sometimes more than a millionfold. The objective was to focus the search down to individual virus particles and to choose those that generated the most offspring.

It was an exercise Dr. Kilbourne and his assistant had

been through many times before. But this time it was tense because it was also a tight race against an unknown competitor—the original swine influenza-type virus that had been found in New Jersey.

Presumably this virus had not formed spontaneously at Fort Dix. It must have come from somewhere. It might have gone somewhere else, but virus specialists searching the nation were unable to find it. Was it breeding anywhere at all outside the laboratory in Manhattan? No one knew. And no one knows today.

But a decision had to be made on whether or not to produce vaccine against the virus, and that decision had to be made well before April 1 or it would be too late for the flu-vaccine manufacturers. Meanwhile, a virus suitable for vaccine production had to be produced.

By the end of the second week, Dr. Kilbourne had such a virus. He labeled it X-53, the latest in a long line of recombinant flu viruses. Those that grew best were always given odd numbers. What he and Miss Pokorny had produced was only a few milliliters in volume—equivalent to only a couple of teaspoonsful—but it was enough for a start. A special messenger took most of the supply of X-53 to Dr. Francis Ennis of the National Institute of Allergy and Infectious Diseases and to the Center for Disease Control during the weekend of February 27.

One drug company was in so great a hurry to get it that they sent a messenger to Dr. Ennis' home in Bethesda, Maryland to pick up a sample.

All four American manufacturers of flu vaccine were working with X-53 by the following week. Samples were also sent abroad. The Soviet Union and China had them by the first week in March.

From that point forward the race moved to the four United States manufacturers of flu vaccine: Merck Sharp &

Dohme, Merrell National, Wyeth Laboratories, and Parke Davis. By working at top speed on the flu virus and on their own production problems, they turned the discovery made in New Jersey in mid-February into a prototype vaccine before the end of March.

When, late that month, President Ford announced a nationwide program to vaccinate the entire population against the new flu virus, he was able to say that the first experimental lots of vaccine had already been produced.

But, recently, the always fickle flu virus seems to have done something for, rather than against humanity.

Dr. Kilbourne said he had discovered a spontaneous mutation—a random unprompted change—in one sample of X-53 that makes it grow much better than any of the other laboratory-engineered recombinants. This new variant also has gone to the manufacturers as another possible aid in the largest vaccination program ever attempted in American history.

The vaccination effort against the so-called swine flu in 1976 proved to be the most ambitious program of its kind in American history and one of the most unfortunate.

What President Ford announced that spring was a campaign to immunize virtually the entire United States population with vaccine paid for by the federal government.

The huge scope of the enterprise and the federal involvement raised liability problems in the minds of the executives of insurance carriers and vaccine manufacturers. These problems were only resolved by an act of Congress after long debate. There were also problems and delays in the manufacture of the huge supplies of vaccine needed. The actual vaccination effort did not begin until October 1, at least two months behind schedule. Only two weeks after the program began there was a nationwide scare because some elderly patients died shortly after receiving the vaccine. The whole program was shut down for several days while scientists investigated. The deaths proved to be simply coincidental and not caused by the vaccination, so far as the investigators could tell, and the immunization program began again.

By mid-December more than forty million Americans had received the vaccine; more than twice as many as in normal years. But another unforeseen problem caused the whole program to be shut down again, this time permanently.

Alert scientists maintaining careful surveillance of this huge newly vaccinated population noticed an unusual cluster of cases of a rare temporary paralysis known as Guillain-Barré syndrome. The timing of these cases linked them clearly to the vaccination.

Guillain-Barré syndrome has been known for many years. Its causes were, and still are, unknown, but the vaccination program of 1976 showed that this terrifying and sometimes fatal paralysis can result from some aspect of the flu vaccination experience.

These fatalities are rare—about one for every two million vaccinations. In the face of an influenza epidemic this would have been an acceptable risk because influenza itself causes about 200 deaths per million cases—400 times the risk of the vaccination—and the danger from flu is even greater among the elderly and others considered at "high risk" because of chronic illness.

The ultimate irony of the 1976 situation, however, was that the huge enterprise in preventive medicine was directed nationwide against an epidemic that never appeared. The scientists who expected it had simply guessed wrong—an outcome they had realized from the start was quite possible. Except for a handful of cases among pig farmers and one or two close contacts, the swine flu virus ignored humans completely all that winter and the next winter too.

In storage freezers in Atlanta there are today millions of doses of swine flu vaccine that may never be used—a frozen monument to the capriciousness of influenza and to the large gaps that still remain in scientists' understanding of this last of the great contagious plagues of mankind.

27

THE KEY TO LEGIONNAIRES' DISEASE

Harold M. Schmeck, Jr.

ATLANTA, Georgia—Under the microscope it appeared to be a bright red cluster of tiny rods nestled inside the much larger purple blob that represented a single animal cell.

"That would have stopped anyone interested in rickettsiae," said Dr. Joseph E. McDade, the scientist of the Center for Disease Control who found the crucial bit of color on a microscope slide a few days after Christmas, 1976.

The slide had been studied months before and the little cluster had been missed, but now the tiny bits of red seemed to jump out of the microscope at him. Dr. McDade was reviewing, one more time, slides prepared the previous August when the legionnaires' disease outbreak in Philadelphia was a medical mystery at a crisis stage. American Legion members who attended a convention in Philadelphia in July were dying of pneumonia. It was an epidemic that might conceivably have been spreading. Public health experts were urgently trying to discover the cause of the disease.

But the search went on all summer and fall and into the winter. The cause was not found. It might have been a virus infection such as swine flu, but that was quickly ruled out. It might have been bacteria, or rickettsia, which are disease-causing microbes smaller than bacteria but larger than viruses, but all the tests proved negative. Many toxic chemicals were suspected and were ruled out for lack of evidence.

The legionnaires' disease remained an unsolved mystery in which about 180 persons had been seriously ill and 29 had died.

That was why, in the week after Christmas, Dr. McDade, a microbiologist at the center's leprosy and rickettsia branch, was again looking at slides prepared the previous summer when a rickettsial disease called Q fever had seemed a likely suspect.

After the holiday, the microbiologist had time to look for subtleties. Furthermore, a detailed report on the long investigation had just been circulated among the scientists at the center in Atlanta. Its conclusion was most unsatisfying: "Cause unknown."

Dr. McDade, Dr. Charles C. Shepard, head of the leprosy and rickettsia branch, and many other specialists at the center were haunted by the feeling that there must be some clue that had been overlooked.

The slides passing under Dr. McDade's microscope held specimens of tissue from guinea pigs that had been inoculated with material from the lungs of persons who had died in the Philadelphia outbreak.

If it was the disease called Q fever, which can produce pneumonia, the guinea pigs would have become ill with fevers and the tiny rod-shaped organisms called rickettsia should have been detectable in their tissues.

In fact, some of the animals had become ill and died, although death was not a common result of the rickettsial infection. But the search for the germs failed.

There was a simple explanation of the animals' deaths and most scientists accepted it at the time. When a person dies, bacteria commonly invade the bloodstream and many tissues. It could have been this kind of post-mortem bacterial invasion that caused fatal infection in the guinea pigs.

In fact, bacteria were found in the guinea pigs, but they were not bacteria of any types or in any pattern that could be linked to legionnaires' disease. For the scientists, it was another dead end.

In the summer everyone had to look fast as well as hard at all the evidence. Although necessary under the circumstances, that hurried atmosphere was not conducive to the best science, said Dr. Shepard, who had been backpacking in Wyoming when the legionnaires' outbreak occurred.

The investigators were at another disadvantage too. Every test they did had to be directed against something specific and each test took time. There simply is no single research test to find "it" in the whole universe of possible causes unless there is some clue to what "it" might be.

Thus, Q fever looked like a possibility, so the scientists tested for rickettsia. Viral pneumonia seemed to fit the picture, more or less, so other experts tested for flu viruses. Each educated guess proved wrong.

The sum total of studies that summer at many laboratories at the Center for Disease Control had largely ruled out known infectious diseases. Much of the emphasis then turned to toxicology, the search for poisonous chemicals ranging from pesticides to the industrial chemical nickel carbonyl. These studies proved to be dead ends, too.

Yet, there had to be a cause, and, for years, the center had prided itself on solving mysteries just like this. Only a few had eluded the center's experts, even though sometimes years elapsed before the solution was found.

Only late in 1976, for example, did experts at the center solve the mystery of a serious epidemic of digestive-tract illness on the island of Truk in the Pacific that had occurred almost thirteen years previously. The island had a population of only 16,000, of whom 5,000 became ill and 7 died. The cause, it turned out, was a virus called a reo-virus that was not discovered until years after the outbreak. The center had kept frozen blood samples from the islanders to test them against newly discovered causes of disease or with new tests that might not have been developed when the samples were put into storage.

One disease outbreak that had never been solved was a pneumonia outbreak at St. Elizabeth's Hospital in Washington in 1965. It bore some intriguing similarities to the legionnaires' disease. The infections all seemed to occur at one time and in one place. There was no outward spread. Many of the Washington victims had died and the epidemic occurred in the summer.

Several scientists had thought of the St. Elizabeth's outbreak when the legionnaires' episode occurred, but there was no way of testing to see if the cause was the same. Some clue had to be found first.

Dr. McDade, a soft-spoken scientist in his mid-thirties, cannot explain exactly why he decided to go back over some of the slides from the legionnaires' investigation except for the post-Christmas lull and the gnawing feeling that things did not quite add up.

Every one of the slides had been looked at before, with about five minutes spent on each. For reasons that, again, he finds difficult to define, the scientist picked one slide for exhaustive study. For a full half hour he sat at the microscope doggedly studying that one slide portion by portion.

"It's like looking for a contact lens on a basketball court with your eyes four inches above the ground," he recalls.

It was during that long, intense study that he found the tiny cluster of rod-shaped objects that appeared bright red because the specimens had been stained with a chemical that would make either rickettsia or bacteria stand out in that color if any were present.

It was an exciting discovery because the rods looked like rickettsia and because their location showed that they had probably been infecting the cell. Dr. McDade showed the slide to Dr. Shepard, an authority on rickettsial disease.

The two scientists agreed it was a clue worth pursuing. They took small samples from the frozen guinea pig tissues, carefully labeled and preserved after the slide had been made, and inoculated these samples into fertilized eggs.

The chick embryos in the eggs died after four to six days and the scientists found many rod-shaped organisms in the tissues.

The hunt was getting exciting, but also puzzling. The rods were far too big to be rickettsia. The scientists thought at first that they were bacteria, but the pace at which they were killing the chick embryos was too slow for that explanation. Furthermore, the rods proved exceedingly difficult to grow on nutrients that common bacteria devour avidly. Whatever the rods were, they were certainly not bacteria of any recognizable species.

The first key question, however, was whether or not they were really linked to the legionnaires' disease. It was entirely possible that they were present, but not the cause. They could be innocent bystanders, so to speak.

To help settle this question the scientists took a preserved blood sample from a legionnaire who had the disease and tested it for antibodies against the new microbe. It was a procedure called a fluorescent antibody test in which the sample would fluoresce if the telltale antibodies were there.

When the test was complete, the scientists were rewarded with a slide that glowed with dramatic confirmation. The scientists had found a bacterialike germ that had been very much at the scene of the crime in Philadelphia. A few more tests confirmed the result and showed that the microbe was absent from persons who had no contact with the outbreak in Pennsylvania.

The research workers needed more legionnaires' blood samples to test. But they had a problem. It was too early to announce that they had found the cause, and if word got out of what they were doing it would probably be big news. Yet they had to have more samples to prove their case. These samples were scarce, extremely valuable commodities.

Dr. Shepard went to the scientist ultimately responsible for the specimens and asked for them, explaining in confidence but in detail why he needed them. The samples were made available at once.

The resultant study linked the newfound microbe solidly to twenty-nine of the thirty-three best-documented cases of legionnaires' disease.

This pretty much ruled out the "innocent bystander" possibility. The animal studies had demonstrated that the bacterialike microbes could cause illness and death. Now these same organisms were found in many cases of well-documented legionnaires' disease, but not in persons who had escaped it. No "innocent bystander" would be found at the scene of so many crimes and nowhere else.

Throughout the entire process of resurrecting clues to a half-year-old mystery, the thoroughness of the center's investigators appeared to have paid off handsomely. Dr. Shepard said the samples on which the studies depended were all properly documented and properly kept and the information and specimens that might be needed to follow up late clues were all available.

This kind of diligence also paid off in another way. The center maintains, frozen for future study, a large archive of blood samples from many sources, including outbreaks of illness that have never been explained. They are stored, row upon row, in little glass bottles with identifying metal seals in freezer lockers that fill an entire large room at the center.

In a sense, it is a frozen catalogue of the center's unfinished business. The blood samples from Truk were kept there as well as the samples from the pneumonia outbreak at St. Elizabeth's Hospital in 1965. There were ninety-four illnesses and sixteen deaths in the epidemic at that federally operated mental institution in Washington.

There were mind-teasing similarities between the St. Elizabeth's epidemic and legionnaires' disease. Having found, at last, bacteria that appeared clearly linked to legionnaires' disease, the scientists had the clue that might link that disease also to the epidemic at the hospital. So they tested the strange legionnaires' microbe against the long-preserved blood samples from St. Elizabeth's. Samples from thirteen of the fourteen cases tested gave strong positive reactions. It appeared clear that the same bacteria that erupted in Philadelphia in 1976 had also been at work eleven years previously at St. Elizabeth's.

Many mysteries concerning the Philadelphia outbreak are still unsolved even though the scientists believe they have found its probable cause.

The microbe they are growing in fertilized eggs remains unrecognized and still behaves strangely for a bacterium. It has been grown only with difficulty on bacterial culture plates. The scientists are not entirely sure that the germ growing in these standard bacterial nutrients is the same as that growing in the eggs.

They do not know where the microbe came from, how it managed to infect 180 men and women in Philadelphia in the summer of 1976, or how often it may have been at the root of other unexplained causes of pneumonia.

But researchers have in their hands an organism that probably caused the disease and was certainly closely linked to it. Having this, they are in a much better position to learn more about the legionnaires' outbreak; how and why it arose and, the scientists hope, what might be done to prevent such outbreaks in the future.

The discovery of the legionnaires' disease bacterium opened up a whole new vista in mankind's understanding of pneumonia. The same microbe was quickly linked not only to the epidemic at St. Elizabeth's Hospital in 1965, but also to a mysterious outbreak in Pontiac, Michigan a few years later, and to several other episodes that had been mysteries for years.

By the fall of 1977 it was clear that the illness nicknamed legionnaires' disease was, and perhaps always had been, one of the natural forms of pneumonia. It was not new but simply undiscovered until the aftermath of the outbreak in Philadelphia. Experts estimate that it is probably the cause of about 2 percent of all unexplained pneumonias of the group doctors call "atypical" and, thus, probably accounts for thousands of serious illnesses every year.

Less than a year after the discovery by Drs. McDade and Shepard, the bacteria had been traced to occasional cases of pneumonia in twenty-nine states and several foreign countries. In the summer of 1977 there were clusters of cases, among other places, in Burlington, Vermont; Columbus, Ohio; and Kingsport, Tennessee.

Just where the bacteria stay in nature and how they sometimes infect humans remains unknown. But at least one antibiotic drug—erythromycin—has been found useful in treating the infections.

An editorial in the *New England Journal of Medicine,* December 1, 1977, praising the work of Drs. McDade and Shepard said: "One can look forward to more rapid accumulation of information about the sources and mode of spread of the organism, the best approaches to treatment of these infections and the possibility that suitable protective vaccines will be prepared in the not too distant future."

28
PENICILLIN-RESISTANT GONORRHEA
Harold M. Schmeck, Jr.

SEATTLE, Washington—He gave the report at a medical meeting in London, predicting that the germ of gonorrhea might abruptly develop total resistance to penicillin, the drug that had been effective against it for decades.

"It was received quite well," Dr. Stanley Falkow recalls. "People thought it was interesting. I don't think one soul there believed me."

The prediction had doomsday overtones, but at that time, 1975, it also seemed unlikely. Gonorrhea is the most common bacterial disease of humans. There are an estimated 100 million cases a year, probably two million of them in the United States. In treating gonorrhea, penicillin has been both effective and cheap. Deprived of its use, venereal-disease programs of some nations might collapse.

Few doctors took the threat of penicillin resistance very seriously. Over the years it took more and more of the drug to kill an infection. A gradual survival-of-the-fittest process was weeding out those strains that were easiest to kill. But the bacteria had never become totally resistant. Why should they change suddenly now?

Dr. Falkow and his group at University of Washington thought it might happen because they had seen just such an abrupt change in another species of bacteria called *Hemophilus influenzae.* It, too, had been a classic case of enduring sensitivity to penicillin and then, abruptly, it was not.

Having studied this change in *Hemophilus* they had sound reasons for thinking it might happen to the gonococcus too. But speculating about it in conversation in the laboratory and writing a formal report using their data on *Hemophilus* as a model for what might happen to the gonococcus were two far different things. The scientists agonized about this before deciding.

The abrupt change in *Hemophilus* was the work of something Dr. Falkow calls "jumping genes." They are little pieces of genetic material that seem almost to hop from one species of bacteria to another, carrying with them capabilities the new owners never had before.

Dr. Falkow had been invited to give a report on antibiotic resistance at a meeting in London on sexually transmitted diseases.

"More than anything," he recalls, "the message I wanted to give them was about these jumping genes. That was new and was something I didn't think was appreciated."

That was the message he gave in London in June 1975 and at later meetings. Scientists admired the science in the report but the prediction seemed farfetched.

Dr. Falkow, a spare, energetic man in his early forties, is a native of Albany, New York. He became fascinated with microbiology in his early teens. As an undergraduate at the University of Maine he took a nonpaying summer job at a hospital to get experience. He has been trying to outguess disease-causing bacteria ever since.

Many convergent trails had led the scientists in Seattle to their prediction—Dr. Falkow's long interest in bacteria and their genetics, his experience with hospital epidemics of drug resistance in the 1950s, another team member's interest in gonorrhea, a new laboratory technique developed by yet another.

They had been studying bacterial genetics—growing the germs in laboratory flasks, exposing them to specific drugs and specific nutrients, then taking the one-celled microbes apart to see what specific pieces of the genetic material DNA (deoxyribonucleic acid) could be matched with specific bacterial traits.

Like many other research groups they were interested particularly in little circular pieces of DNA called plasmids. These were separate from the cell's main genetic machinery, yet were capable of giving the bacteria instructions that would pass from generation to generation. It was as part of plasmids that the "jumping genes" moved from one species of bacteria to another.

It was a natural recombinant DNA process—like the deliberate experiments called gene-splicing, that have aroused controversy in recent years. Bacteria knew how to do these tricks naturally long before scientists figured out how to do them experimentally. The gene that Dr. Falkow's team found most interesting was one that carried instructions for making an enzyme called penicillinase. This enzyme chews up the drug penicillin and makes it harmless.

Once a plasmid carries into a bacterium the gene for making that enzyme, there is little need for survival-of-the-fittest selection. The bacteria will automatically become totally resistant to penicillin even if they have had no previous exposure to the drug.

Dr. Falkow and his colleagues saw that happen with *Hemophilus influenzae,* common bacteria that sometimes cause serious illness in children. There was an outbreak of meningitis in Washington, D.C. caused by *Hemophilus.* Surprisingly, the germs were not touched by penicillin's modern close relative ampicillin.

"As soon as we heard that, we just knew they had plasmids," Dr. Falkow recalls. They got samples of the bacteria and went to work; growing the germs in liquid nutrients and then adding a detergent to split the cells open in a gentle way that allows the plasmid DNA to float in the liquid above the solid parts of the disrupted bacterial cells.

Then, through a complex process that employs some of the techniques of recombinant DNA research, they found the plasmids and proved that these did indeed carry the gene for making penicillinase.

Lyn Elwell, a member of the team, developed a way of sorting out the DNA fragments from a cell so rapidly that he could analyze more than a dozen different samples of bacteria in a day. He let each sample migrate through a slab of special gelatinous material under the influence of electric current in a simple device of his own design. An experienced worker could identify the plasmids from the pattern of bands in the gel made visible by their pinkish glow under fluorescent light.

The group screened several hundred strains of the bacteria to learn their traits.

Meanwhile, another member of the group, Leonard Mayer, had been studying the gonococcus for entirely different reasons and found that these bacteria too had plasmids. The gonococci did not have the gene for making penicillinase on any of their plasmids, but, in theory, it was reasonable that they might pick it up just as *Hemophilis* did.

A worrisome part of that possibility was that, once the *Hemophilus* had picked up the trait for resistance to penicillin, other types of antibiotic resistance seemed to flow into it readily. The germs developed resistance to tetracycline, kannamycin, and other drugs too. It was as though an important barrier had been breached, letting many things in, Dr. Falkow said.

The group's prediction was based on all of their painfully gathered threads of research. Meanwhile, the gonococcus stayed helpfully sensitive to penicillin and the world went on as usual.

Then, last fall, Dr. Falkow got a telephone call from the Center for Disease Control in Atlanta, the federal facility that keeps watch for the United States on infectious diseases all over the world.

Public health officers in the Philippines were finding gonorrhea cases totally resistant to penicillin, the scientist in Seattle was told. Surprisingly the same thing had been reported almost simultaneously from England. The center sent Dr. Falkow specimens of the bacteria. These arrived on a Thursday. By Sunday the research workers knew that the two resistant gonococci had different plasmids, but both provided the ability to manufacture penicillinase and fight off any attack by penicillin.

Marylin Roberts, a member of the research team who had worked extensively with the plasmids of *Hemophilus influenzae,* said that these looked a lot like the material from the Far East. She tested the newly arrived samples.

But soon she ran into a problem. The team needed large supplies of the plasmids for further studies. She tried to introduce them into laboratory strains of gonococci in the conventional way—adding the plasmids themselves to a bacterial culture of gonococci. It did not work. Time after time the experiment was tried, but each time it failed. The transfer worked in nature, but it would not work in the Health Sciences Building of the University of Washington Medical School.

This was both a roadblock and a puzzle that needed to be solved. How did the bacteria do it naturally?

It turned out that the transfer worked if a bacterium containing the plasmid was put in physical contact with one that did not. For bacteria this constitutes sexual transfer of the genetic material. Miss Roberts brought it about in the gonococci by carefully plating out on a piece of filter paper enough bacteria so that cells containing the plasmids touched those that did not.

"It seems ironic," another scientist said later, "that the gonococci have a sex life of their own."

Meanwhile, scientists all over the world have been watching the gonococci apprehensively to see how far the drug resistance will spread. By now, resistant strains have appeared almost everywhere, including the United States, where specimens have been found from coast to coast.

Fortunately, these still account for only a fraction of all gonorrhea. For these cases the new, but expensive, antibiotic spectinomycin is effective. For the rest, penicillin still works and remains the drug of choice.

It appears that any plasmid puts an extra metabolic burden on a bacterium, Dr. Falkow explains. For that reason plasmids may persist in only a minority of the bacteria unless something forces them to become universal. Dr. Falkow, for years a critic of indiscriminate use of antibiotics, notes that the gonococci may all become resistant to penicillin only if mankind forces them to do so by haphazard and therefore ineffective use of the drugs.

But he also thinks that the last has not been heard of the jumping genes for penicillinase.

One interesting puzzle, still to be solved, is where did that particular gene come from in the first place? Did *Hemophilus* give it to the gonococcus? Did each of these bacteria get it from another common source? If so, which one?

Today no one knows, but Dr. Falkow is willing to offer one more prediction.

What disease-producing germ is likely to get the gene for penicillinase next, he was asked?

The meningococcus, the scientist answered without a moment's hesitation. This is a major cause of bacterial meningitis, an often-deadly brain infection that can develop with frightening speed. If this germ does follow the example of *Hemophilus influenzae* and the gonococcus, it could create yet another crisis of drug resistance for the world.

29

A NEW METHOD OF INTRAVENOUS FEEDING SAVES LIVES

Jane E. Brody

HOUSTON, Texas—The operations were technically successful, but despite the best available hospital care all three patients died, and a very discouraged young intern, Dr. Stanley J. Dudrick, began to question his desire to become a surgeon.

But instead of changing careers that weekend in Philadelphia sixteen years ago, Dr. Dudrick decided to look into the reasons for the deaths. The explanation he came up with, severe malnutrition, led him to develop a way to feed people entirely by vein, a technique that can help save the lives or vastly improve the health of perhaps one in five hospital patients.

Many seriously ill or injured persons could recover completely if they did not starve to death first. Previously, only one-quarter or less of a patient's necessary nourishment could be given intravenously.

Largely through Dr. Dudrick's work, the medical profession over the last decade has grown increasingly aware of a previously unrecognized problem in American hospitals—the fact that many patients die not because their illness is incurable but rather because they are unable to eat or absorb enough nutrients to sustain life.

Dr. Dudrick and others estimate that in at least 10 percent

and perhaps as many as 30 percent of deaths that occur in hospitals malnutrition is the direct or an important contributing cause. There are no reliable statistics because malnutrition is rarely listed as a cause of death; rather, the death certificate cites cancer, kidney failure, multiple injuries, or whatever.

The growing recognition by the medical profession of the extent of this problem and the gradual adoption of Dr. Dudrick's technique to cope with it exemplifies how medical knowledge expands and develops into new ways to save lives. As more and more doctors use the new method, they have found that many patients whom they would otherwise expect to die can survive and recover if they are properly nourished.

But still only half the hospitals use the technique, and malnutrition remains a stubborn problem that proves fatal to thousands each year.

Too often, Dr. Dudrick said, hospital patients receive superb care but they die of "hospital malnutrition" because the doctors think of everything except nutrition. He added, "It's senseless to pay $200 to $300 a day for intensive care and then not pay the extra $50 to $100 for nutrition. It's like buying a Mercedes without any wheels on it."

Although only a decade old, the technique of total intravenous nutrition that Dr. Dudrick developed has already revolutionized the care of otherwise doomed patients in thousands of hospitals around the country. For some disorders, instead of 80 percent of patients dying or requiring mutilating operations, 80 percent survive and get well with little or no surgery.

In analyzing the three deaths that occurred while he was on call at the University of Pennsylvania Medical Center in 1961, Dr. Dudrick found that the patients, in the course of their illnesses, had become malnourished to the point that they were simply unable to recover.

He talked the problem over with his professor and mentor,

Dr. Jonathan Rhoads at the University of Pennsylvania, where unlike other medical centers there was a long tradition of interest in nutrition. Dr. Dudrick quickly realized that death from malnutrition happened in hospitals everywhere every day—to patients with severe bowel disease, cancer, extensive burns, multiple injuries, and kidney and liver disease, as well as to accident victims in comas and babies born prematurely or with severe defects of the digestive tract.

Surely, the young doctor figured, there must be some way to nourish people who cannot get enough food by the usual route. Feeding through a tube directly into the stomach or small intestine was adequate for some, but it was little help to patients whose digestive tracts were not working properly or who required "supernutrition" to keep them alive.

And the traditional intravenous (IV) feeding of sugar water was clearly inadequate for more than a few days of supplemental nutrition. At most, it could supply 500 to 600 calories a day, and even if fortified with vitamins and minerals and alternated with a solution of predigested protein, it was not enough to sustain body weight or prevent the breakdown of tissues in the average adult or growing child.

Dr. Dudrick, who is now chairman of the department of surgery at the new University of Texas Medical Center, explained that conventional intravenous nutrients are dripped into a small vein, usually in the arm, at a concentration of 5 percent, which matches the normal concentration of dissolved substances in the blood.

If the nutrient concentration in the IV drip is increased much beyond that, the veins and blood cells might be damaged. And if more than three bottles of the dilute solution, each containing about 175 calories, are given in one day, all the water would seriously impair the patient's heart and lung function.

There were other obstacles to completely nourishing a person by vein, Dr. Dudrick recalled. There was at best a poor understanding of exactly what nutrients to supply and in what amounts to maximize their utilization and minimize the chances of toxic side effects. There was the danger of blood clots and infection associated with establishing a long-term connection between the sterile bloodstream and the hostile, microbe-laden world outside.

And most inhibiting of all, he said, were the many pronouncements over the past two or three decades from prominent physicians and scientists that total nutrition by vein was "either impossible, or at best improbable and impractical."

But Dr. Dudrick, who describes himself as "a stubborn Pole," was not deterred. Over the next two years, in thirty-seven patients who faced imminent death from malnutrition, he dared to increase the concentration of the IV solution to 10 and even 15 percent and, by giving diuretics simultaneously to rid the body of excess water, he also increased the number of bottles of IV solution to four or five and sometimes as many as seven a day.

"This was the horse-and-buggy stage—very primitive—but we showed that potentially it would help," Dr. Dudrick said. "Many of the patients stopped losing weight. But they didn't gain weight either, and we realized that we were going to have to get in still more nutrients if we wanted to get such people well."

So Dr. Dudrick took his concept back to the laboratory where, working with beagle puppies, he devised the "Model-T Ford" of total intravenous nutrition. He reasoned that if the small veins could not withstand a concentrated nutrient solution, then why not go to a bigger vein where the greater blood flow would rapidly dilute the intravenous meal. He chose the superior vena cava, which returns blood from the head and neck to the heart, and can be entered through the chest.

Using beagle puppy littermates and an IV solution with a nutrient concentration of 30 percent, he showed that the dogs would grow and develop normally even if their entire small intestine were removed and they were fed solely by vein twenty-four hours a day.

"This was the first time anyone had grown any animal entirely by vein, and it encouraged us to try again in patients," Dr. Dudrick recounted. "In the next year, six patients we expected to die because of inadequate nourishment agreed to be fed by the new technique. Two did die in spite of it, but four left the hospital really well."

Between 1964 and 1966, Dr. Dudrick and his coworkers repeatedly demonstrated the life-sustaining value of the feeding technique in hundreds of research animals. In man, he figured that by giving three liters of the 30 percent solution, an amount that most patients could tolerate without diuretics, he could supply an adult with 2,500 to 3,000 calories of dextrose (a simple sugar), amino acids (the building blocks of protein), electrolytes like sodium and potassium, vitamins and minerals, occasionally supplemented by essential fats.

Such predigested nutrients could enter cells directly from the blood and sustain life and growth even if no food at all were taken by mouth. And the exact mix of nutrients could be adjusted according to the needs of individual patients, with some solutions containing as many as thirty-seven different components.

Then in 1967, Dr. Dudrick's new technique, which he calls total parenteral nutrition or intravenous hyperalimentation (IVH), faced its most challenging trial. A baby was born with a deformity of the bowel that would have been fatal. Dr. Dudrick began feeding the infant by vein, and the scrawny 4-pound baby started to fill out and grow. After forty-five days on IVH the baby weighed 7.5 pounds and had grown 2 inches, and she continued to thrive on vein feeding for twenty-two months.

In eighteen consecutive infants with bowel deformities

treated this way at the University of Pennsylvania, there were no deaths, and as the infants' remaining guts became sufficiently developed or the deformity corrected itself, they were weaned from the intravenous feeding and placed on ordinary food.

Then Dr. Dudrick faced what he calls his "first true adult nutritional emergency." She was a fifty-two-year-old woman, five feet two inches tall, who following stomach surgery had gradually dropped to forty-nine pounds and was practically moribund.

"She was dying right before our eyes because she was out of gas—no fuel," Dr. Dudrick observed. "We plugged her in, giving her only 500 or 600 calories a day at first because you can actually hurt starving people by trying to feed them too much too fast, and gradually increasing to 3,000 calories a day by IVH.

"In two months, she was up to 79 pounds and a month later she was sent home eating normally and weighing 100 pounds," he said. "She's been between 104 and 110 ever since. A true nutritional save. And nutrition was all the treatment she got—no medicine and no fancy operation."

Dr. Dudrick said he has been publicly derided by some surgeons for saying that in many cases, surgery represents "an admission that you have failed to control the condition medically and had to resort to a mutilating operation." Nonetheless, he added, "I tell all my surgical residents that if they are worth their salt, they will spend their time finding ways to put themselves out of business."

Dr. Dudrick has spared hundreds of patients from having to go under the knife by nourishing them intravenously until their condition cleared up spontaneously. For example, in a series of fifty-two patients with severe inflammatory bowel disease (ileitis and colitis) that could not be controlled by drugs and who now faced surgical removal of their diseased bowel, more than half recovered spontaneously just with intravenous nutritional therapy—no drugs and no surgery.

In another quarter of patients, the disease also cleared up and the bowel was saved, but surgery was needed to remove an intestinal blockage or to close a hole that would not heal. Only one in five patients had to have a part of his bowel removed to relieve his condition.

Dr. Dudrick and his colleagues have also found that instead of surgery, which often fails to correct the problem, nutritional therapy alone could produce spontaneous healing in 70 percent of cases of gastrointestinal fistulas, abnormal openings in the intestinal tract that cause leakage of food, wastes, and body fluids and can lead to life-threatening infection and malnutrition.

"Bowel disease," Dr. Dudrick said, "is a case of double jeopardy. You need as much nutrition to heal a sick bowel as you would need to fight disease anywhere else in the body. But you're unable to get in the nutrients you need to heal with.

"And if you feed a diseased bowel, you abrade inflamed lesions with food, bile, feces, pancreatic juice, and acid. You wouldn't think of putting anything like that on an open wound on the skin. But with intravenous nutrition, we can put the bowel at total rest so that it can be restored to normal faster."

Most of Dr. Dudrick's patients are able to return to work or useful life. "These people are sharp of mind and otherwise healthy of body," he noted. "They just don't have a way to get the nutrients they need by the gastrointestinal tract."

One forty-two-year-old man who does not have enough functioning bowel to live on ordinary food alone was fitted with a portable IVH unit through which he is nourished twenty-four hours a day. He mixes his own nutrient solutions under aseptic conditions at home and periodically comes to the hospital for tests. Meanwhile, he works eight to twelve hours a day as a trucker, earning $35,000 a year.

But the majority of IVH patients require only temporary nutritional support. Among them are victims of severe burns and multiple injuries, who require enormous amounts of calories and protein while they recover. Because of the destruction of tissues and the body's protein-consuming response to injury, a burn victim may need as many as 10,000 calories a day, Dr. Dudrick said, and there is no way to supply anything like that amount without IVH.

Similarly, cancer patients may become temporarily malnourished from the side effects of treatment, which can cause loss of appetite, vomiting, and diarrhea. Dr. Dudrick has found that when they are supported by IVH, cancer patients can withstand considerably higher doses of anticancer drugs, thus possibly increasing their chances for cure.

There also appears to be a slightly better response to ordinary levels of cancer therapy when patients get IVH, possibly due to the improvement in immunological response that Dr. Dudrick and others have observed among IVH patients.

Perhaps Dr. Dudrick's most unusual case of malnutrition was a man weighing 600 pounds. The man had so much fat in his blood in relation to protein that his 18-inch incision following a hernia operation would not heal until he was placed on high-protein IVH.

All patients receiving IVH must be carefully monitored to be sure they obtain the right balance of nutrients and electrolytes. At first, daily tests are made of the patients' blood and urine, with adjustments made in the nutrient mix until stable desired levels are obtained.

Currently, IVH costs from $30 to $50 a day for the raw materials, plus up to $100 a day for studies and labor, plus the cost of hospitalization.

Dr. Dudrick confesses to never having had a formal course

in nutrition, a topic that is usually included in biochemistry courses in medical school. He said he learned most of what he knows from reading and talking with others knowledgeable about nutrition.

In the process, he has evolved a new way of looking at the entire subject. "I no longer think of food as ice cream sundaes, steak, pork chops, or the eggs and juice you have for breakfast," he said. "Nutrition really is the delivery of adequate chemical substances to the cells from the blood. It's what gets into the cells that counts."

Today, one in six patients at Dr. Dudrick's hospital is on IVH and he admits that when he first started feeding beagles by vein, he had "no concept of what this was going to lead to. No way in my wildest imagination would I have thought of all the reasons for which IVH is now being used."

But now he is thinking ahead. His department, which he founded five years ago at the age of thirty-six (making him the youngest chairman of surgery in the country at the time), maintains fifteen research laboratory rooms for biochemical and immunological studies and an animal research facility.

His plans include using IVH to look for special diets that may preferentially starve cancer cells or enhance their susceptibility to therapy, and trying to isolate the precise dietary components that promote clogging of the arteries and heart disease.

As "a driven man, totally dedicated to medicine," he gives little thought to cutting back on the sixteen-hour days he typically puts in at the medical center six days a week. (He tries to keep Sunday a family day to spend with his wife—whom he describes as "a wonderful supportive lady"—and six children, aged twelve to eighteen.)

Currently, Dr. Dudrick estimates, about half the nation's hospitals are using IVH at least some of the time. But a lot more who could use it are not, he said.

"IVH is a team effort. It takes a whole new investment, an educational process, a lot of hard work and dedication, and some people just don't want to be bothered with the hard work," he remarked.

Still, he added, "I visualize the day when IVH will be routine, when doctors wouldn't not think of treating every patient with adequate nutrition. To be consistently good rather than flash-in-the-pan, medicine still has to be practiced with the basics, and nutrition is basic to us all."

BIOGRAPHIES OF AUTHORS

JOHN NOBLE WILFORD is director of science news of *The New York Times.* Born in Kentucky, he received his B.S. from the University of Tennessee and M.A. from Syracuse University and spent a year of further study at Columbia University. He worked as a reporter on the *Commercial Appeal* in Memphis and the *Wall Street Journal* and was a contributing editor of *Time.* He joined *The Times* in 1965 and covered all the major missions of space exploration, including the moon landings. He is author of *We Reach the Moon,* the first definitive account of the Apollo Project after the Apollo 11 moon landing in July 1969. His reporting has also taken him through the eye of a hurricane to get a story on cloud seeding, to the floor of the Grand Canyon with a mapping party, to the Soviet Union (where he became the first Western correspondent to visit the Soviet space center), to Greenland with the International Ice Patrol, and to Scotland with an expedition searching for the Loch Ness Monster.

WALTER SULLIVAN, science editor of *The New York Times,* joined the newspaper as a copy boy in 1940, immediately following his graduation from Yale University. During World War II, he commanded a destroyer in the Pacific. He rejoined *The Times* in 1946, this time as a reporter, serving in China, Korea, and Berlin. In 1946–47, 1954, 1956, and again in 1976 he covered Antarctic expeditions. Mr. Sullivan became science editor in 1966. He is a fellow of the American Association for the Advancement of Science, the Arctic Institute of North America, and Jonathan Edwards College at Yale. He is a member of the American Geophysical Union and the council of the American Geographical Society. In 1959 Mr. Sullivan won the George Polk Memorial Award for his "distinguished coverage of the International Geophysical Year." He has also received the Westinghouse/AAAS award for science writing, the James T. Grady Award of the American Chemical Society, the American Institute of Physics–U.S. Steel Foundation Award in Physics and Astronomy, the Bradford Washburn Award of the Boston Museum of Science, and the Daly Medal of the American Geographical Society. His books include: *Quest for a Continent, Assault on the Unknown, We Are Not Alone, Continents in Motion,* and *Black Holes: The Edge of Space, The End of Time.*

BAYARD WEBSTER, assistant director of *The New York Times* science news department, writes on environmental, wildlife, and nature subjects in addition to general science. Born in East Orange, New Jersey, he attended Wesleyan University and Johns Hopkins University. After college he was for a time a tennis pro, partner in a wholesale coffee business, and copy writer for a Manhattan advertising agency. In World War II he was a Navy fighter pilot in the Pacific. After the war he became a newspaper reporter, working for the *Greenwich* (Conn.) *Time* and *Baltimore Sun,* where he was a Pentagon correspondent, editorial writer, and acting city editor. When he joined *The Times* in 1957, Mr. Webster became an Albany correspondent, assistant city editor, and environmental writer.

LAWRENCE K. ALTMAN is one of the few medical doctors working as a full-time daily newspaper reporter. Born in Quincy, Massachusetts, he went from Milton Academy to Harvard, from which he graduated in 1958. He received his M.D. from Tufts University School of Medicine in 1962. Dr. Altman's internship was at Mt. Zion Hospital in San Francisco. Then he served for three years in the U.S. Public Health Service as editor of its *Morbidity and Mortality Weekly Report,* a journal dealing with reported cases of communicable diseases in the world. From 1966 to 1968 he was a resident in internal medicine at the University of Washington, Seattle. He became a member of *The New York Times* science news department in 1969. His reporting honors include the Claude Bernard Science Journalism Award in 1971 and 1974. He is also a clinical assistant professor at the New York University Medical School.

JANE E. BRODY is a medical reporter for *The New York Times* and author of the weekly "Personal Health" column. She was born in Brooklyn, majored in biochemistry at Cornell University (B.S., 1962), and studied journalism at the University of Wisconsin (M.S., 1963). She also spent one summer as a laboratory technician at the Sloan-Kettering Institute in New York and another doing biochemical research under a National Science Foundation fellowship. Ms. Brody was a reporter on *The Minneapolis Tribune* before joining *The Times* in 1965. She has won the Howard Blakeslee Award for articles on heart disease and cholesterol; the Sigma Delta Chi Deadline Club award for coauthorship, with Boyce Rensberger, of a five-part series on incompetent medical practices; and the J.C. Penney-University of Missouri Newspaper Award for consumer affairs reporting for her "Personal Health" column. She is author of *You Can Fight Cancer and Win,* published in 1977.

MALCOLM W. BROWNE became a science reporter for *The New York Times* in 1977 after a distinguished career as a foreign correspondent in Europe, Asia, and Latin America. He was born in New York City and educated at Swarthmore College, New York University, and Columbia University. For five years, before entering journalism, he was an industrial chemist. Mr. Browne's first newspaper job was as reporter and editor with *The Middletown* (N.Y.) *Daily Record.* In 1964, he shared a Pulitzer Prize for his reporting from South Vietnam for the Associated Press. In addition, he was an Edward R. Murrow Fellow of the Council on Foreign Relations and won the World Press Photo Grand Prize, an Overseas Press Club award, the George Polk Memorial Award, a Louis M. Lyons Award, and a Sigma Delta Chi Award. Mr. Browne is the author of *The New Face of War.*

VICTOR K. MCELHENY, who joined *The New York Times* in 1973, was born in Boston and received a bachelor's degree in sociology from Harvard College. Later, he was a Nieman Fellow at Harvard, European correspondent for *Science,* and science editor of *The Boston Globe.* He has also served as science commentator for WGBH-TV in Boston and was a columnist for *Technology Review* at the Massachusetts Institute of Technology. In 1974, he wrote the first newspaper article about successful use of so-called "gene splicing" to transfer genes from animal cells into single-cell bacteria. In 1976, he inaugurated the weekly "Technology" column for *The Times* financial section. In May, 1978, he left *The Times* to become director of the Banbury Conference Centre at Cold Spring Harbor Laboratory.

BOYCE RENSBERGER began writing about science at the University of Miami, where he trained as a marine biologist before switching to journalism. With a bachelor's degree in zoology and journalism (1964) he went to Syracuse University under a National Institute of Mental Health fellowship for a master's degree in writing about psychiatry and behavioral sciences. In 1966 he joined the *Detroit Free Press* as a science and medical writer. There he was responsible for special sections on space flight (coinciding with the first manned lunar landing) and on environmental/ecological issues. In 1971 Mr. Rensberger joined *The Times* as a science writer, leaving in 1973 to accept an Alicia Patterson Foundation fellowship to spend a year in Africa researching man's evolutionary past and ecological present. In 1974 he rejoined *The Times.* Major areas he has covered include archeology, the world food situation, human evolution, and the quality of medical practice. He is the author of *Cult of the Wild,* published in 1977. He was born in Indianapolis.

HAROLD M. SCHMECK, JR., specializes in the coverage of biology for *The New York Times.* He is a graduate of Cornell University and was a 1953 Nieman Fellow at Harvard University. He served as an Air Force navigator in World War II. Before joining the staff of *The Times* he was a reporter for the Danville, Illinois *Commer-*

cial-News and the *Rochester Times-Union,* where he also wrote a science column. He has covered science for *The Times* for twenty-one years, eleven of them as science correspondent in Washington. In addition, he has written magazine articles for a broad range of publications, radio scripts for Voice of America and CBS Radio Network, and two books—*The Semi-Artificial Man* and *Immunology, the Many-Edged Sword.* He won a Sigma Delta Chi Deadline Club Award for science writing in 1970.

INDEX

A, G, C, and T (nucleotides), 122–123, 126
Aboriginal Americans, 36–42
Academy of Natural Sciences, 181, 182, 184
Adaptive radiation phenomenon, 23–24
Adovasio, James M., 36–42
African army worm, pest-control research, 138
Air pollution, 12
Alpha particles, 55
Altman, Lawrence K., 149–154, 256–257
American Chemical Society, 217
American Telephone and Telegraph Company, 80
Ames Research Center, 134
Amino acid analyzer, 31
Amino acid racemization, 30–31, 34
Amino acids, 120, 132, 248
Ancient Bristlecone Pine Forest, 15
Anderson, Lennart, 65
Annealing, DNA strands, 125
Anolis (genus), 145
Anopheles mosquito, 146
Antimatter (positrons), 52
Antineutrons, 57
Applied Science Center of Archeology, 15
Arteriosclerosis, 215
ATCase molecule, 155, 156–159
ATG (adenine, thymine, and guanine), 121–122
ATP synthesis, 133
Australopithecus, 23
Autopsies, on mummies, 7–12
A/Victoria (virus), 221, 222

Baboquivari Mountain, 70
Bacteriorhodospin, 132–134
Bada, Dr. Jeffrey, 28, 30–31, 32, 34, 35
Baker, Tom, 99, 101
Barraco, Dr. Robin, 10
Barrell, Bart G., 119
Bear's winter sleep pattern, 149–154
Beckes, Mike, 39
Bell, Alexander Graham, 79
Bell Telephone Laboratories, 79–86
Bering land bridge, 27
Bhabhas, 53, 54, 58
Blaurock, Dr. Allen E., 132
Bog People, The (Glob), 203
Bormann, Becky, 175
Bormann, Dr. F. Herbert, 174–179
Boyce Thompson Plant Institute, 108
Brady, Dr. Roscoe, 213–220
Bristlecone pine, 13–19
Broca's area, 21, 24, 25
Brody, Jane E., 95–102, 110–116, 129–136, 161–166, 201–206, 244–253, 257
Brookhaven National Laboratory, 8, 49, 60
Brooks, Chester E., 82
Browne, Malcolm W., 155–160, 257–258
Bruinsma, Dr. Oebele H., 141–142
Burghardt, Dr. George M., 147

Calhoun, Dr. John B., 198–199
California Institute of Technology, 66
California State University, 28
Cambridge University, 119
Cardé, Ring, 101
Cardiovascular Research Unit (University of California), 131
Carotene pigments, 131
Cartifacts, 35
Cary Arboretum, 103, 105
Cascade Range, 62
Carter, Dr. George, 34–35
Center for Biomedical Research, 162–163
Center for Disease Control, 222, 225, 229–236, 241
Center Magazine, 51
Centrex telephones, 84
CERN (European atomic center), 55
Charm particles, 50
Chaudhury, Dr. Mohammed, 139–140
Chlorophyll, 130
Chromosome barrier, 104
Cline, Dr. David B., 56, 59
Clovis spear points, 33
Cockburn, Dr. Aidan, 9, 10
Cockburn, Eve, 9
Columbia University, 21, 85
Cornell University, 133, 155–156, 175
Cosmic rays, 54
Crick, Dr. Francis H. C., 118–119
"Crick strand," 123, 124
Cro-Magnon Man, 28, 34

Darwin, Charles, 206
Davis, Dr. David E., 199
Deconvolution process, 73
Del Mar skull, 32, 35
Dendrochronology, 13–19
Deoxyribonucleic acid (DNA), 117–128, 156, 239, 240
Detroit Institute of Arts, 7
Diatoms and streams, research on, 180–185
Dichlorodiphenyl-trichloroethane (DDT), 102
Djerassi, Carl, 139
Double helix, 118–119
Double Helix, The (Watson), 119
Drell, Dr. Sidney D., 52
Dudrick, Dr. Stanley J., 244–253
Dutch elm disease, 103

"East Africa: Science for Development" (Odhiambo), 138
Eastern Timber Wolf Recovery Team, 191, 192
Eaton, Dr. Gordon P., 62–67
Ecological Society of America, 174
Edison, Thomas Alva, 79
Einstein, Albert, 52
Electroantennography, 96, 98
Electron-positron pairs, 53, 54
Electrons, 52
Electrophoresis process, 125
Elm trees, breeding, 103–109
Endocasts, 22–24
Energy Research and Development Administration, 55
Enewetak rats, 193–197
Ennis, Dr. Francis, 225
Entomological Society of America, 96
Enzyme replacement, medical treatment, 213–220
Enzymes, 155–160, 214
Escherichia coli, 120
Evans, Jean, 160
Experiment 1-A (India), 56–57
Exxon Foundation, 146, 147

Fabry's disease, 214, 216
Falkow, Dr. Stanley, 237–243
Fawcett, Dr. Don W., 164

Federal Communications Commission, 84
Feiner, Alec, 79–86
Ferguson, Dr. C. Wesley, 15–16, 17
Fermi National Accelerator Laboratory (Fermilab), 50, 55–56, 59
Ferreed crosspoint, 79–86
Fiddes, Dr. John C., 118, 120
Finnegan, Dr. Michael, 10
Folsom spear points, 33
F-1 plants, 111, 112, 113
Forbes, John, 64
Ford, Gerald, 222, 226
Ford, Dr. William T., 58
Forest watershed ecosystems, 173–179
Fossil skulls, 20–26
Foundations of Archeology (Smith), 28
F-2 plants, 111–112
Fucosidosis, 214

Gal, Dr. Andrew E., 219
Galactic images, 74
Galaxies, infancy of universe and, 68–75
Gamma rays, 52
Gaucher's disease, 214, 216, 217–220
Gene-splicing, 239
G-4 virus, 120, 121
Gibbs Laboratory, 155, 159, 160
Gilbert, A. C., 159
Glob, P. V., 203
Glucocerebrosidase, 219
Godson, Dr. C. Nigel, 118, 121
Goldfield, Dr. Martin, 221–222
Gonorrhea, penicillin-resistant, 237–243
Gonzaga University, 58
Goodman, Dr. Howard M., 118
Gorham, Dr. Eville, 201–206
Greene, Dr. Harry W., 147
Guerrero, Stella, 144–148
Guillain-Barré syndrome, 227
Gunn, Dr. Joel, 40

Halbedel, William, 73
Halobacterium's purple membrane, 129–136
Harari, Dr. Haim, 50
Harvard University, 55, 122, 155, 164
Hawaiian Volcano Observatory, 61, 67
Hawaii Volcanoes National Park, 61
Heart disease, 12
Heizer, Dr. Robert, 34
Hemophilus influenzae, 237–243
Henderson, Richard, 136
Herodotus, 11
Hibernation, bear's, 149–154
Hodge, Dr. Charles, 184
Hoffman, Dr. Raoul, 155–156
Holloway, Dr. Ralph L., Jr., 20–26
Homo erectus, 26
Homos, small-brained, 23
"Hospital malnutrition," 245
Hubbard Brook Experimental Forest, 173, 175
Hubble, Edwin P., 71

Ice Age, 33
Imlay, Dr. Richard, 58
Insect pest control, 95–102
Insect research, 137–143
International Centre of Insect Physiology and Ecology (I.C.I.P.E.), 137–143
Intravenous feeding, 244–253
Intravenous hyperalimentation (IVH), 248–253
Iron Age, 203
Isle Royale National Park, 187

Jackson, Dr. William, 193–197, 199

Jaggar, Dr. Thomas A., 62–63, 66
Jensen, Bradley, 58
John and Alice Tyler Ecology Award, 184
Johns Hopkins Medical School, 9

Kanfer, Dr. Julian, 219
Kansas State University, 10
Kaons, 57
Karnosky, Dr. David, 103–109
Kau Desert, 64
Kennedy, Dr. Gail, 29, 31–32, 34
Kennedy Space Center, 59
Ketchledge, Raymond W., 82
Kety, Dr. Seymour, 216
Kilauea crater, 63–66
Kilbourne, Dr. Edwin, 221–226
Kirkpatrick, Ralph, 71
Kitt Park National Observatory, 69
Known Age Project, 15
Kochansky, Jan, 101
Kolar gold mines, 56
Korean War, 216

Labor, division of, 40
Laboratory of Molecular Biology, 117–118, 122
Laboratory of Tree Ring Research, 15
La Jolla pits, 28, 29, 30, 35
Lamberson, Phillip, 195
Lanyi, Dr. Janos, 134
Leakey, Richard, 20
Leeuwenhoek, Antony van, 164
Legionnaires' disease, 229–236
Lewisville site (Texas), 33–34
Likens, Dr. Gene E., 175, 179
Ling, Dr. T. Y., 58
Lipids, 214, 215–216
Lipid storage diseases, 220
Lipscomb, William Nunn, Jr., 155–160
Lovell, Clarence A., 82, 83
Lowry, T. N., 83
Lunate sulcus, 22–23
Lynds, C. Roger, 68–75
Lynn, Dr. George, 9–10

McDade, Dr. Joseph E., 229–236
MacDonald, Dr. Russell E., 134
McElheny, Victor K., 79–86, 258
MacNeish, Dr. Richard S., 32, 33
Maise stem borer, pest-control research, 138
Malaria research, 144–148
Male contraceptive, 161–166
Mann, Dr. Alfred K., 56
Man's antiquity (in the Americas), 27–35
Martin, Dr. Paul, 33
Massachusetts Institute of Technology, 60, 63
Mauna Loa volcano, 61–65
Max Planck Institute, 98
Mayo Clinic, 149–154
Meadowcroft Rockshelter excavation, 36–42
Mech, Dr. L. David, 186–192
Medical Research Council (Great Britain), 122
Memorial University (Canada), 51
Meningitis, bacterial, 242–243
Merck Sharp & Dohme Company, 225–226
Merrell National Company, 226
Metachromatic leukodystrophy, 214, 216
Methuselah tree, 15
Michael, Dr. Henry N., 13–19
Milky Way, 70, 71
Miller, Albert, 41
Millner Canyon, alluvial fans, 18–19
Minus method, DNA nucleotides, 127–128
Morse, Samuel F. B., 79
Mosquito, pest-control research, 138

Mount Palomar, 69
Mt. Sinai School of Medicine, 221, 223
Mummy autopsies, 9–12
Muons, 53, 54, 58

National Aeronautics and Space Administration (NASA), 134, 135
National Institute of Allergy and Infectious Diseases, 225
National Institute of Neurological Diseases and Blindness, 216
National Institutes of Health, 131, 135, 198, 214, 216, 219
National Museums of Kenya, 20, 24
National Radio Astronomy Observatory, 69
National Science Foundation, 22, 175
Nelson, Dr. Ralph A., 149–154
Neolithic Age, 17
Neutrinos, 56–57
New England Journal of Medicine, 236
New York Botanical Garden, 105
New York State Agricultural Experiment Station, 95, 110
New York State College of Agriculture and Life Sciences, 133
Niemann-Pick disease, 216
Norway rats, 193
Nucleotides, 120, 121–122

Obenchain, Dr. Fred, 140
Odhaimbo, Thomas R., 138–139, 142–143
Oesterhelt, Dr. Dieter, 132–133
Oloo, Dr. Gilbert W., 141
Orr, Dr. Peter, 58
Owens River quarry, 19

Paleopathology Association, 8–9, 12
Panofsky, Wolfgang Kurt Herman, 49
Papago Indians, 69, 70
Parke Davis Company, 226
Particles, subatomic, 49–60
Patrick, Frank, 182
Patrick, Dr. Ruth, 180–185
Pauling, Linus, 157
Peabody Foundation for Archeology, 32
Peatlands, 201–206
Peck, Bill, 7
Penner, James, 151
Pennsylvania Power and Light Company, 184
Pennsylvania State University, 199
Pentchev, Dr. Peter G., 219
Perutz, Dr. Max F., 119
Pest control, use of pheromones for, 95–102
Petrosian, Dr. Vahé, 71
Pheromones, 95–102
Phillips, Dr. David M., 161–166
Phi X-174 (virus), 117, 118, 119–120, 121
Phrenology, 21
Pikimachay Cave, 32
Pine Alpha tree, 15
Pions, 57
Plasmodium balli, 146
Plasmodium falciparum, 146
Plasmodium tropidari, 146
Plus method, DNA nucleotides, 128
Pokorny, Barbara, 222, 224
Polar coordinate steroplotter, 25
Polynesian rats, 195–196, 200
Poole, Kathy, 99
Population Council, 162–163
PR-8 (flu virus), 223–224
Progestin (hormone), 161
Proteins, 120
Protons, 60
Psi (or J) particle, 49, 50–51, 60
Purdue University, 113

Q fever, 230, 231

Racker, Dr. Ephraim, 133, 135
Radcliffe College, 160
Radioactive carbon 14 dating, 13–14, 19, 30, 33–34, 205
Rand, Dr. A. Stanley, 144–148
Rats, 193–200
Rensberger, Boyce, 7–12, 20–26, 27–35, 36–42, 137–143, 186–192, 258–259
Reyman, Dr. Theodore A., 10, 11
Rhoads, Dr. Jonathan, 246
Rhodopsin, 132
Ribonucleic acid (RNA), 118
Richter, Dr. Burton, 51, 54, 60
Rick, Dr. C. M., 114–115
Ridinger, P. G., 83
Roberts, Marylin, 241–242
Robinson, Dr. Michael H. and Mrs., 148
Robinson, Dr. Richard W., 110–116
Rockefeller University, 130, 131, 163
Roelofs, Dr. Wendell, 95–102
Roundworm infestation, 11
Rowen, Robert, 131
Royal Society of London, 164
Rubbia, Dr. Carlo, 55
Rutgers University, 184
Rutherford, Sir Ernest, 54–55

St. Elizabeth's Hospital, pneumonia outbreak (1965), 232, 234–235, 236
Sandage, Dr. Allan, 72, 75
Sandia Laboratories, 64
Sanger, Dr. Frederick, 118–19, 121, 122
Schistosomiasis infestation, 11
Schmeck, Harold M., Jr., 213–220, 221–228, 229–236, 237–243, 259
Schneider, Dietrich, 98
Schwitters, Roy F., 53
Science (magazine), 34, 138, 147
Scott, Paul A., 71
Scripps Institution of Oceanography, 31, 32
Shapiro, Dr. David, 219
Shepard, Dr. Charles C., 230–236
Siberia, stone tools, 38
Silver Canyon, alluvial fans, 18–19
Smith, Dr. Jason W., 28–30, 35
Smithsonian Institution, 144
Sorghum shootfly, pest-control research, 138
Sphagnum (peat moss), 202, 203–204, 205
Sphingosine, 216
Standard candle brightness, 72
Stanford Linear Accelerator (SLAC), 49, 50–54, 56, 57, 59–60
Stanford Research Institute, 18
Stanford University, 71, 139
Stoeckenius, Dr. Walther, 129–136
Stone-Age hunters, 36–42
Stone tools, 29, 38
Stroud Water Research Center, 180–182, 184
Suess, Dr. H. E., 17
Sullivan, Walter, 49–60, 68–75, 117–128, 144–148, 255–256
Superior National Forest, 187
Swine flu vaccine, 221–228

Tapeworm infestation, 11
Tay-Sachs disease, 214, 216
Technical University, 85
Telephone technology, 79–86
Temple Unversity, 15
Termites, pest-control research, 138, 141–142
Testosterone (hormone), 162
Texas A&M University, 35
Ticks, pest-control research, 138, 140–141
Ting, Samuel C. C., 60

Tollund Fen (peat bog), 203
Tomatoes, frost-resistant, 110–116
Trams, Dr. Eberhard, 218
Tree-ring dating (dendrochronology), 13–19
Tropical Research Institute (Smithsonian Institution), 144
Truk Island, digestive-tract illness (1976), 231–232
Tsetse fly, pest-control research, 138, 139
Twain, Mark, 63, 64
Twinkle, star's, 73–74

U.S. Fish and Wildlife Service, 187, 191
U.S. Forest Service, 175, 176
U.S. Geological Survey, 61, 62, 67
U.S. Public Health Service, 177, 198, 222
Universidad del Valle, 146
University of Arizona, 15, 16, 17, 33, 41
University of California at Berkeley, 22, 34
University of California at Davis, 114
University of California at Los Angeles, 29
University of California at San Diego, 15, 17
University of California at San Francisco, 118, 129
University of Colorado, 9
University of Illinois, 9
University of Maine, 238
University of Pennsylvania, 10, 15, 17, 56, 58, 181, 249
University of Pennsylvania Museum, 8, 17
University of Pittsburgh, 37
University of Tennessee, 147
University of Texas, 8, 40, 246
University of Virginia, 182
University of Washington, 237, 241
University of Wisconsin, 56, 58, 105
Unwin, Nigel, 136

Vessey, Dr. Stephen H., 194–200
Volcano predictions, 61–67

Wadhams, Frances, 98–99
Walker, Alan, 25
Wanderer, Dr. Peter, 58
Water pollution research, 180–185
Watson, Dr. James D., 118–119
"Watson strand," 123, 124
Watt, James, 82
Wayne State University, 7, 10
Webster, Bayard, 103–109, 173–179, 180–185, 256
Weissman, Dr. Sherman M., 121
Weizmann Institute, 50
Wellik, Dianne, 150–151
Wesleyan University, 65
White Mountain National Forest, 173–179
White Mountains (California), 13–16, 17
Whitney, Eli, 79
Wilford, John Noble, 13–19, 61, 255
Wolves, 186–192
World War I, 204
Wyeth Laboratories, 226

X-53 (virus), 225, 226

Yaes, Dr. Robert, 51
Yale School of Forestry and Environmental Studies, 174
Yale University, 117, 118, 121, 175, 178

Zimmerman, Dr. Michael, 10
Zollman, Dr. Paul E., 151–152